Lecture Notes
in Business Information Processing 290

Series Editors

Wil M.P. van der Aalst
Eindhoven Technical University, Eindhoven, The Netherlands
John Mylopoulos
University of Trento, Trento, Italy
Michael Rosemann
Queensland University of Technology, Brisbane, QLD, Australia
Michael J. Shaw
University of Illinois, Urbana-Champaign, IL, USA
Clemens Szyperski
Microsoft Research, Redmond, WA, USA

More information about this series at http://www.springer.com/series/7911

Rim Jallouli · Osmar R. Zaïane
Mohamed Anis Bach Tobji · Rym Srarfi Tabbane
Anton Nijholt (Eds.)

Digital Economy

Emerging Technologies and Business Innovation

Second International Conference, ICDEc 2017
Sidi Bou Said, Tunisia, May 4–6, 2017
Proceedings

Springer

Editors

Rim Jallouli
University of Manouba
Manouba
Tunisia

Rym Srarfi Tabbane
University of Manouba
Manouba
Tunisia

Osmar R. Zaïane
University of Alberta
Edmonton, AB
Canada

Anton Nijholt
University of Twente
Enschede
The Netherlands

Mohamed Anis Bach Tobji
University of Tunis
Tunis
Tunisia

ISSN 1865-1348   ISSN 1865-1356 (electronic)
Lecture Notes in Business Information Processing
ISBN 978-3-319-62736-6   ISBN 978-3-319-62737-3 (eBook)
DOI 10.1007/978-3-319-62737-3

Library of Congress Control Number: 2017946062

Printed on acid-free paper

This Springer imprint is published by Springer Nature
The registered company is Springer International Publishing AG
The registered company address is: Gewerbestrasse 11, 6330 Cham, Switzerland

Preface

The Second International Conference on Digital Economy, ICDEc 2017, took place in the lovely white and blue village of Sidi Bou Said, Tunisia, one of the most visited places around the Mediterranean Sea, and recognized as one of UNESCO's world heritage sites under the name of "Carthage-Sidi Bou Said." We met to celebrate and discuss digital economy.

The theme of ICDEc 2017 was "Digital Economy: Emerging Technologies and Business Innovation." The conference offered a number of sessions discussing innovative research focusing on emerging technologies which support the digital transformation of business and the economy, with a particular emphasis in this edition on data science and security, machine learning, Web data, cloud computing, smart cities, digital marketing, and e-learning.

All papers submitted to the conference were reviewed using a double-blind peer-review process. Papers that needed major revision went through a second round of review. We received a total of 46 qualified papers. At the end of the reviewing process, 18 full papers were accepted for presentation at ICDEc 2017. The number of reviewers per paper varied between three and six, with an exact average of 3.7 reviews per paper. All accepted papers were presented at the conference.

We express our appreciation to everyone who contributed to achieve the objectives of the ICDEc 2017 project: An international conference with participants and partners from more than 30 countries. Our sincere appreciation goes to the ICDEc keynote speakers for presenting and discussing the new trends of the digital revolution: the ethics of big data; the mass adoption of cloud computing and Web technology; precision medicine with machine learning; smart cities; and social media and value creation.

We would like to express our deepest gratitude to the conference program chairs for their effort and expertise. We would like to thank the country chairs, the Organizing and Finance Committees, as well as the Scientific and Program Committees for their support in making this conference successful. Special thanks go to the Higher School of Digital Economy, ESEN University of Manouba, Tunisia, for supporting the organization of ICDEc 2017.

We would also like to thank the sponsors of the conference for their contribution toward helping the Tunisian Association of Digital Economy ATEN to achieve its goal of the annual International Conference on Digital Economy ICDEc.

June 2017

Rim Jallouli
Osmar Zaiane

Organization

General Co-chairs

Rim Jallouli   University of Manouba, Tunisia
Osmar R. Zaiane   University of Alberta, Canada

Program Committee Co-chairs

Mohamed Anis Bach Tobji   University of Tunis, Tunisia
Anton Nijholt   University of Twente, The Netherlands
Rym Srarfi Tabbane   University of Manouba, Tunisia

Special Session Co-chairs

Chiheb Eddine Ben N'cir   University of Tunis, Tunisia
Yamen Koubâa   France Business School, France

Organizing Committee Chair

Jihène El Ouakdi   University of Manouba, Tunisia

Organizing Committee

Salima Abbes   ISET'COM, Tunisia
Hamida Amdouni   University of Manouba, Tunisia
Houda Challakhi   University of Manouba, Tunisia
Dhouha Doghri   University of Manouba, Tunisia
Nabila El Jed   University of Manouba, Tunisia
Dorra Guermazi   University of Manouba, Tunisia
Ines Mezghani   ISET'COM, Tunisia
Lamia Zaibi   University of Manouba, Tunisia

IT Chair

Nassim Bahri   One Way IT, Tunisia

Finance Co-chairs

Afef Herelli   University of Tunis-El Manar, Tunisia
Karim Kammoun   University of Manouba, Tunisia

Junior Committee

Fatima Chaouachi	University of Manouba, Tunisia
Mahmoud Ghandour	University of Manouba, Tunisia
Chayma Maatougui	University of Manouba, Tunisia
Rihab Melki	ISET'COM, Tunisia

Country Chairs

Ana Pires	Federal University of Bahia, Brazil
Chiheb El Ouakdi	Laval University, Canada
Abdel-Badeeh Salem	Ain Shams University, Egypt
Yamen Koubâa	Brest Business School, France
Masayuki Maruyama	Kansai University of International Studies, Japan
Dyah Ismoyowati	Universitas Gadjah Mada, Indonesia
Ali Afshar	Eqbal Lahoori Institute of Higher Education, Iran
Kaouther Znaidi	College of Business Administration, University of Hail, KSA
Mohammad Makki	School of Business - Lebanese International University, Lebanon
Dorota Jelonek	Czestochowa University of Technology, Poland
Rute Abreu	Instituto Politécnico da Guarda, Portugal
Sambil Charles Mukwakungu	University of Johannesburg, South Africa
Che-Jen Su	Fu Jen University, Taiwan
Walid Trabelsi	IBM Ireland, UK
Samir R. Moussalli	Huntingdon College, USA

Scientific Committee

Taymoor Abdelgaber	Ain Shams University, Egypt
Dagmar Caganova	Slovak University of Technology, Slovakia
Akil Elkamel	Northern Borders University, KSA
Mohamed Imen Gallali	University of Manouba, Tunisia
Atef Gharbi	NBU, KSA
Sihem Guemara El Fatmi	SUP'Com Tunis, Tunisia
Arpan Kar	Institute of Technology Delhi, India
Mohamed Limam	Dhofar University, Oman
Mohammad Makki	Lebanese International University, Lebanon
Olfa Nasraoui	University of Louisville, USA
Evgeny Nikulchev	Technological Institute, Russia
Posegga Joachim	University of Passau, Germany
Maria Reznakova	Brno University of Technology, Czech Republic
Mornay Roberts-Lombard	University of Johannesburg, South Africa

Mohamed Roushdy	Ain Shams University, Egypt
Abdel-Badeeh Salem	Ain Shams University, Egypt
Nadine Sinno	Lebanese International University, Lebanon
Lorraine Warren	Massey University, New Zealand
Dominique Wolff	IRSI ESC, France
Ezzeddine Zagrouba	Virtual University of Tunis, Tunisia

Program Committee

Ryma Abassi	ISET'COM, Tunisia
Mohamed Amine Abid	University of Passau, Germany
Ali Afshar	Eqbal Lahoori Institute of Higher Education, Iran
Mona Al-Achkar Jabbour	Lebanese Information Technology Association, Lebanon
Zeyad Alfawar	University of Dammam, KSA
Paulo Almeida	Leiria Polytechnic, Portugal
Raouia Ayachi	University of Tunis, Tunisia
Mohamed Karim Azib	University of Tunis El Manar, Tunisia
Noor Azlinna Azizan	University Malaysia Pahang, Malaysia
Deny Bélisle	Université de Sherbrooke, Canada
Rafika Ben Guirat	American University in the Emirates, United Arab Emirates
Mohamed Ben Halima	University of Sfax, Tunisia
Sonia Ben Slimane	Novancia Business School, France
Afef Ben Youssef	ISET'COM, Tunisia
Waâd Bouaguel	University of Carthage, Tunisia
Imen Boukhris	LARODEC, University of Tunis, Tunisia
Zaki Brahmi	University of Manouba, Tunisia
Manuel Castro	Universidad Nacional de Educación a Distancia, Spain
Mouna Chebbah	University of Tunis, Tunisia
Soumaya Cheikhrouhou	Université de Sherbrooke, Canada
Mohamed Dbouk	Lebanese University, Lebanon
Amira Eleuch	Brest Business School, France
Chiheb El Ouakdi	Laval University, Canada
Hédia El Ourabi	École de Sciences de Gestion, Université du Québec, Canada
Yamen El Touati	Northern Border University, KSA
Amira Essaid	University of Tunis, Tunisia
Yamna Ettarres	University of Manouba, Tunisia
Tahani Gazdar	University of Manouba, Tunisia
Houda Hakim Guermazi	University of Manouba, Tunisia
Wided Guezguez	Umm Al Qura University, KSA
Ahlem Hajjem	École de Sciences de Gestion, Université du Québec, Canada

Manel Hamouda	University of Gabes, Tunisia
Abdallah Handoura	Telecom Bretagne, France
Farah Harrathi	University of Manouba, Tunisia
Jamel Henchiri	University of Gabes, Tunisia
Reaan Immelman	Barclays Africa, South Africa
Dyah Ismoyowati	Universitas Gadjah Mada, Indonesia
Naila Khan	Birmingham City University, UK
Gaurav Khatwani	Indian Institute of Management, India
Petros Kostagiolas	Ionian University, Greece
Cao Lanlan	NEOMA Business School, France
Kun Chang Lee	Sungkyunkwan University, South Korea
Krassimir Markov	Institute of Information Theories and Applications, Bulgaria
Olfa Mannai	University of Tunis El Manar, Tunisia
Hamid Mcheick	Université du Québec, Canada
Samir Moussalli	Huntingdon College, USA
Mercy Mpinganjira	University of Johannesburg, South Africa
Klimis Ntalianis	Athens University of Applied Sciences, Greece
Nesrine Omrani	Paris School of Business, France
Sevim Oztimurlenk	Long Island University, USA
Malgorzata Pankowska	University of Economics in Katowice, Poland
Angela Pereira	Politécnico de Leiria, Portugal
Célia Rafael	Polytechnic Institute of Leiria, Portugal
Rhouma Rhouma	University of Manouba, Tunisia
Ahmed Samet	Université Rennes 1, France
Jeanne Schreurs	Hasselt University, Belgium
Aymen Sioud	Université du Québec, Canada
Yassine Slama	University of Manouba, Tunisia
Layth Slimane	EFREI, France
Anna Soltysik-Piorunkiewicz	University of Economics in Katowice, Poland
Mourad Touzani	NEOMA Business School, France
Imene Trabelsi Trigui	University of Sfax, Tunisia
Sulov Vladimir	University of Economics, Bulgaria
Widyawan Widyawan	Universitas Gadjah Mada, Indonesia

Additional Reviewers

Reema Aswani	Jaypee University of Information Technology, India
Dorra Attiaoui	Université Rennes 1, France
Hajer Bellalouna	University of Manouba, Tunisia
Amal Ben Rjab	Université Laval, Canada
Meriam Belkhir	University of Sfax, Tunisia
Ikbel Daly	University of Manouba, Tunisia
Fatma Ezzahra Bousnina	University of Tunis, Tunisia
Sayda Elmi	University of Tunis, Tunisia

Mohamed Aymen Haj Kacem	University of Tunis, Tunisia
Siwar Jendoubi	Université Rennes 1, France
Nimish Joseph	Institute of Technology Delhi, India
Safa Kaâbi	University of Manouba, Tunisia
Anis Lachiheb	University of Sousse, Tunisia
Badran Raddaoui	Université de Poitiers, France
Hans P. Reiser	University of Passau, Germany
Ines Thabet	University of Manouba, Tunisia
Asma Trabelsi	University of Tunis, Tunisia
Feriel Zerzeri	University of Manouba, Tunisia
Kaouther Znaidi	University of Hail, KSA

Organizer

Association Tunisienne d'Économie Numérique

Scientific Partners

École Supérieure d'Économie Numérique

University of Manouba

Alberta Machine Intelligence Institute

   Laboratoire de Recherche Operationnelle,
de Decision et de Controle des procedes

   Tunisia Chapter of the IEEE Computational
Intelligence Society

Sponsors

      Bureau des Études Techniques d'Assistance
et de Pilotage

BETAPi
Bureau des Etudes Techniques
d'Assistance et de Pilotage

   WESS E-COMMERCE

      Bourse de Tunis

   Arab Tunisian Bank

      Office Plast

Media Partners

   Challenges TN

 Borderline Creatives

 Information Technology Mag

 Plumes Économiques

 Réalités

 Tunivisions

Contents

Digital Marketing

Celebrity Endorsement on Social Networks Sites: Impact of His/Her Credibility and Congruence with the Endorsed Product, on the Consumer's Information Adoption and Dissemination

Nadia Ben Halima[1], Hamida Skandrani[2(✉)], and Nawel Ayadi[3]

[1] IHEC, Carthage, Tunisia
[2] ISCAE, LIGUE, Manouba, Tunisia
hamida.skandrani@gmail.com
[3] ISG, Tunis, Tunisia

Abstract. This study aims at examining the impact of celebrities' credibility and congruence with the endorsed product on the consumers' information adoption and dissemination on Social Networks Sites (SNS). Two different and spontaneous celebrities' endorsements were considered. The first is more cause-related and concerned the celebrity Hend Sabry, a well-known actress, who endorsed on Facebook the national campaign "Zourou Tounes" ("visit Tunisia"), in order to promote the tourist destination Tunisia after the terrorist attack at the Bardo National Museum (18th of March, 2015). The second is more product/brand-related and involved the celebrity Dorra Zarrouk, a notorious actress, who endorsed on Instagram the chocolate product of a well-known brand. 168 Tunisian Internet users participated to the online survey after visiting the Facebook post of each endorsement. The findings revealed differences in the effect of celebrity's credibility on consumers' information adoption and dissemination. In particular, for Hend Sabry's endorsement, only the celebrity's expertise dimension of credibility has an effect on the adoption of the information spread about the destination Tunisia by the followers. For Dorra Zarrouk's endorsement, only the celebrity's reliability dimension of credibility plays an important role in the consumers' adoption and dissemination of online information. The results show also that the perceived congruence between the celebrity and the product positively influences the information adoption and its dissemination by consumers.

Keywords: Celebrities' endorsement effectiveness · Credibility · Congruence · Social networks sites (SNS) · Cause-related endorsement · Product/brand endorsement

1 Introduction

Celebrity endorsement has been and still represents a very important and attractive research topic for marketing academicians and practitioners. Such interest seems to be

© Springer International Publishing AG 2017
R. Jallouli et al. (Eds.): ICDEc 2017, LNBIP 290, pp. 3–14, 2017.
DOI: 10.1007/978-3-319-62737-3_1

even more significant with the high penetration rates of social networks site (SNS)[1]. Indeed, celebrities are recognized to be particularly influential within the consumer-buying process [1] in so far as they are considered as a powerful and trustworthy information source [2–5]. Relying on the opportunities those celebrities may provide to their products/brands [1], firms are increasingly investing billions of dollars[2] (off-line and online) to build partnership with these famous personalities and, consequently, to achieve superior values in terms of consumer's perception of products and brands and buying intentions [6]. In 2015, Forbes[3] reported that celebrity endorsements continue to increase yearly, with an estimated \$50 billion being spent globally on traditional forms of advertising (television, magazines and advertising posters, etc.) but also celebrity pages on social networks [7].

Despite the recognition of celebrity endorsement as one of the most used and effective advertisement strategies [8, 9], and the attractiveness of SNS offering people the opportunity to engage with whom they want, including, friends, favorite brands, fans or celebrities [10], celebrity endorsements in the SNS environment and its antecedents need to be more thoroughly addressed [7]. The investigation into celebrity endorsement's effectiveness appears to be meaningful for at least three reasons. First, given the already shown differences between communication models for traditional and interactive media [7, 11–13], the transferability of effective communication strategies to online platforms and particularly to SNS requires more investigation. Second, as stated by Metzger and Flanagin [14] (p. 211), digital media have provided access to an unprecedented amount of information available for public consumption. Consequently, finding the information that meets user's needs throughout a large number of information providers and identifying the most reliable information becomes critical issues [14].

Third, even though the use of celebrities on SNS allows to circumvent the problems of advertising congestion and consumers' advertising avoidance [15], there is a lack of studies guiding marketers' actions in emerging social media environments [7] (p. 143). More specifically, considering different celebrities with diverse online spontaneous endorsements (product/service/cause) and taking into account the credibility and congruence of the information source with the endorsed product/cause, are likely to contribute at a better understanding of the outcomes of the celebrity's endorsement on online environments, in general, and on SNS, in particular. Therefore, this study aims to assess the impact of celebrities' perceived credibility and congruence with the endorsed product on the consumers' information adoption and dissemination on SNS.

In the following, we first present the theoretical background and the main research hypotheses. Next, we expose the adopted methodology. Then, the results are highlighted

---

[1] For example, in 2016, Facebook increased its penetration to 89% of US internet users, whilst Facebook-owned Instagram came 2nd with 32% penetration, according to Global social media research summary 2017 (http://www.smartinsights.com/social-media-marketing/social-media-strategy/new-global-social-media-research/).

[2] http://www.nicekicks.com/nike-spent-almost-10-billion-endorsements-2016/.

[3] https://www.forbes.com/sites/pauljankowski/2015/03/02/9-steps-to-creating-successful-brandcelebrity-partnerships/#6d82f6c21ed6.

and discussed. Finally, the conclusion, study limitations and further research avenues are presented.

2   Theoretical Background and Hypotheses

Celebrity endorsement is defined as a process by which a celebrity expresses his/her opinion about a product, service, or destination. Often related to an activity that he/she is paid for, the endorsement can be made "for free" [16]. Previous studies have shown that celebrities draw attention to the product [17, 18], stimulate brand recall [19, 20], increase firm's profit [21], boost positive product evaluations [22–24], brand credibility, and consumer's buying intensions [2, 22, 25, 26].

The effectiveness of celebrity endorsement and its determinants have been widely investigated [20–29]. The credibility of the celebrity (information source) and his/her congruence with the product, are among the most important antecedents of the endorsement effectiveness in offline and/or online settings, including SNS [26, 30–32]. Yet, some authors called for more studies on online celebrity endorsement taking into account moderating or mediating effects of other factors related to the endorser [33], to the category of the endorsed product/service [7, 33], or to the information receiver [7, 14, 33, 34] etc. These variables may alter the expected effect of the celebrity endorsement in terms of consumer adoption and dissemination of the information he/she spreads about the endorsed product.

2.1   Celebrity Credibility

The credibility of an information source, defined as: "a communicator's positive characteristics that affect the receiver's acceptance of a message" (Ohanian [35], (p. 41), is a subjective variable based on a judgment made by the receiver of the message [36]. According to Erdogan [8], information from a credible source can influence believes, opinions, attitudes and/or behaviors. Metzger and Flanagin [14] rationalize the impact of source credibility in an online environment based on the liking/agreement heuristic (well documented in studies of social cognition and persuasion), which argues that people often agree with those they like.

According to the endorsement literature [37] the components of source credibility are source expertise, trustworthiness [27, 28] and attractiveness [35]. Trustworthiness refers to the source (endorser) perceived objectivity and honesty with regard to the provided information [25]. Expertise is defined as the extent to which the endorser is perceived as skilled, experienced and having a certain level of knowledge with the product category [27]. Attractiveness refers to how physically attractive, elegant, or likable the source is to the audience [8, 25, 33].

Previous studies results showed a positive impact of celebrity's perceived credibility on consumers' positive attitudinal and behavioral responses to the endorsed brand and the product in offline or online environments [7, 8, 25, 33, 34]. It is quite surprising to notice that very few recent research works investigated celebrity endorsement within the on SNS context [38, 39]. Wood and Burkhalter [7] argue that celebrity appeal is especially important to consider in SNS, as approximately one-third of Americans follow celebrities online [40]. A survey by the Economist (2016) shows that such influencers with large online followers are heavily paid for the brand endorsement inasmuch they allow companies to reach a vast network of potential customers[4]. This is likely to help in drawing attention and disseminating brand information [7].

Besides, given that the information adoption and dissemination by consumer stand for attitudinal and behavioral responses too [41], and that the same three-dimensional structure of the celebrity credibility construct (trustworthiness, expertise and attractiveness) was also used in previous studies on online celebrity endorsement, including SNS [16, 42, 43], the following hypotheses and sub-hypotheses are stated:

H1:   The celebrity credibility has a positive effect on the consumer's information adoption in the SNS context

- H1-1. The perceived celebrity's trustworthiness has a positive effect on consumer's information adoption
- H1-2. The perceived celebrity's expertise has a positive effect on consumer's information adoption
- H1-3. The perceived celebrity's attractiveness has a positive effect on consumer's information adoption

H2:   The perceived celebrity's credibility has a positive effect on the consumer's information dissemination in the SNS context.

- H2-1. The perceived celebrity's trustworthiness has a positive effect on consumer's dissemination of the information
- H2-2. The perceived celebrity's expertise has a positive effect on consumer's dissemination of the information
- H2-3. The perceived celebrity's attractiveness has a positive effect on consumer's dissemination of the information

2.2   The Perceived Congruence Between Celebrity and Product

Earlier studies [44] and recent research [45] advocated that the endorser congruence with the product or the brand influences consumer behavior and triggers positive attitudes toward ads and brands. Congruence is identified as a bi-dimensional construct involving: relevance and expectancy [46]. Relevance refers to the degree to which stimulus information allows for a clear identification of the communicated information, or on the contrary prevents it [9]. The expectancy refers to the degree to which information falls within a predetermined pattern or structure evoked by that theme [9]. Indeed, in the endorsement literature, fit or congruence typically refers to the similarity or consistency between the brand and the celebrity [26]. The 'match-up hypothesis' suggests that endorsers are more effective when there is a

---

[4] Someone with 3 m-7 m followers can charge, on average, $187,500 for a post on YouTube, $93,750 for a post on Facebook and $75,000 for a post on Instagram or Snapchat, (http://www.economist.com/blogs/graphicdetail/2016/10/daily-chart-9).

good 'fit' between the endorser and the product (Kamins 1990 [47]; cited in [26]). This match between both the celebrity's and the brand's images has a significant effect on consumer's information adoption and dissemination (i.e. recommendation) on SNS [48] and on brand beliefs and attitudes [33]. This leads to suggest the following hypotheses and sub-hypotheses:

H3: The perceived celebrity's congruence with the endorsed product has a positive effect on the consumer's adoption of the information in the SNS context.

- H3-1: the more the celebrity is considered relevant to testimony on the quality of the product, the more is the consumer's adoption of the information
- H3-2: the more the celebrity is expected to testimony on the quality of the product, the more is the consumer's adoption of the information

H4: The perceived celebrity's congruence with the product has a positive effect on the consumer's dissemination of the information on SNS.

- H4-1: the more the celebrity is considered relevant to endorse the product, the more likely is the consumer's information adoption
- H4-2: The more the celebrity is expected to endorse the product, the more is the consumer's information on SNS.

The following conceptual framework was developed (Fig. 1).

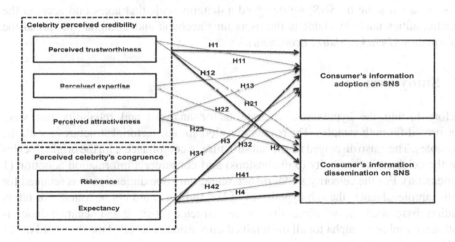

Fig. 1. The proposed conceptual framework

3 Methodology

As stated before, this study aims to examine the impact of celebrities' credibility and his/her congruence with the product on consumer's information adoption and dissemination. Thus, we conducted an online survey using two Tunisian celebrities' endorsements (a cause-related and a product/brand-related ones) through real advertising campaigns on two different SNS (Facebook and Instagram). The followers' number and a pre-test among 30 students allowed identifying the two most perceived notorious celebrities, namely Hend Sabry and Dorra Zarrouk, who perform in Egypt and the Arabic World. Participants were invited to list the popular Tunisian celebrities who can influence their buying decisions.

To reduce the biases related to the commercial character of the endorsement and the possible use of aggressive marketing tactics to enhance the endorsement impact by firms, we considered only endorsements that are *a priori* spontaneous. Hend Sabry (1.2 million followers) endorsed on Facebook the campaign "visit Tunisia", in order to promote the tourist destination Tunisia after the terrorist attack at the Bardo National Museum (18th of March, 2015). This endorsement is more cause-related. Dorra Zarrouk (3 millions followers) endorsed the chocolate product of a well-known brand on Instagram. A convenience sample of 168 Tunisian Internet users (84 for each endorsement) was used in this study. The respondents were aged between 18 and 54 years old and 53.57% of them are females and 42.85% were students. Participants were asked to complete two separate questionnaires after visiting the page of each endorsement.

The measurement scales (7-points) were adopted from previous research on online celebrities' endorsement except for the information dissemination one. Celebrity's credibility and congruence with the endorsed product/cause were respectively measured using Ohanian's [34] and Fleck-Dousteyssier et al. [49] scales. For information adoption, we adopted the Wu and Shaffer scale's [50] 4 items scale. As to the information's dissemination scale on SNS, we designed a 4 items scale that takes into account the functionalities made available to the users on Facebook and Instagram to spread the information (*I click; I share; I comment*).

4   Study Results

Before testing the hypotheses, separate factor analyses and reliability test were performed for both samples. The results revealed similar factorial structures for all the variables. They also displayed a factorial structure in compliance with the original scales for the celebrity's credibility (3 dimensions) and consumer's information adoption (1 dimension). For the celebrity's congruence scale, only one dimension was retained for both samples. Hence, the sub-hypotheses of H3 and H4 could not be tested. For information dissemination, two dimensions were extracted (positive and negative dissemination). Cronbach's alpha for all the retained dimensions exceeded the threshold of 0,7 [51] in the two samples.

Given the lack of a sound theoretical support underlying the proposed conceptual links, a regression analysis is deemed appropriate for hypotheses testing. To check for the normality assumption, we used the Skewness and Kurtosis statistics. In Dorra Zarrouk study, the values of these two coefficients ranged from 0.822 to 0.125, and from −1.379 to 0.732, respectively for Skewness and Kurtosis coefficients. Only the attractiveness dimension displayed a skewness and kurtosis values respectively of −2,345 and 5,325 were slightly higher than the generally accepted level (3). Nonetheless, the standard error for the two coefficients exhibited levels between −1 and 1, indicating a non departure of the normality assumption[5]. It should be noted that overall, the skewness and kurtosis values are below those considered by Kline [52] as problematic (3.0 and 8.0, respectively). The data from Hend Sabry's study diplayed skewness and Kurtosis values meeting the recommended levels.

The Variance Inflation Factors (VIF) was also computed in order to test for the multicollinearity effect. All VIF values, were less than 4 for the two data sets (specifically between 1.534 and 3.236). Thus, multicollinearity, is not a concern as VIF values are lower than 10 [53]. we examined the quality of the regression model via the Fisher-Snedecor test. The R square (R^2) and the standardized coefficients (β) values allowed valuing the potency and the nature of the associations between the variables (i.e. celebrity's trustworthiness, expertise and attractiveness, celebrity's congruence, consumer's information adoption and dissemination). More details about the study results follow.

4.1 Study 1: Hend Sabry's "Visit Tunisia" Campaign

The regression analysis results for Hend Sabry's endorsement revealed that only the celebrity's perceived expertise has a positive effect on consumer's information adoption ($\beta = 0.267$; $p = 0.022$). Hence, only hypothesis H1-2 is empirically supported. No empirical evidence was found for the impact of Hend Sabry's perceived credibility on her followers' dissemination of the information about the campaign "visit Tunisia" ($p > 0.05$ for all the credibility dimensions). H2 was than not supported. However, results showed that Hend Sabry's perceived congruence with the endorsed cause "visit Tunisia" significantly influences her followers' information adoption on Facebook ($\beta = +0.637$; $p = 0.000$), giving empirical evidence to hypothesis H3. The effect of celebrity's perceived congruence with the endorsed cause on the consumer's information dissemination was also supported by the data ($\beta = 0.542$; $p = 0.000$). H4 is than supported.

4.2 Study 2: Dorra Zarrouk's Chocolate Campaign

The regression analysis results showed that Dorra Zarrouks's perceived credibility impact on her followers' information adoption is somehow different from Hend sabry's one. Indeed, while the Dorra Zarrouks's perceived expertise had no significant effect on this adoption ($p = 0,248$), her perceived trustworthiness was found to have a positive effect on consumer's information adoption. Nonetheless, unlike expected results, this celebrity's attractiveness was shown to influence negatively their followers' information adoption on SNS ($\beta = -0.263$; $p = 0.050$). Hence only H1-1 was empirically supported. Besides, only Dorra Zarrouks's perceived trustworthiness had a positive and significant impact on consumer's information dissemination on Instagram, allowing us to support H2-1.

Likewise, Dorra Zarrouks's perceived congruence with the chocolate significantly affects her followers' adoption of information on Instagram (+0.682; $p = 0.000$) and his information dissemination ($\beta = 0,633$; $p = 0,000$). Therefore H3 and H4 are supported

---

[5] http://spss.espaceweb.usherbrooke.ca/pages/stat-descriptives/procedure-explorer.php.

by data. It should be noted that our results are consistent with Kamins and Gupta's [23] ones.

5 Theoretical and Managerial Implications

Overall, partial empirical evidence was found for the study hypotheses. Our results are in line with previous researches that have shown that the extent, to which consumers perceive a source as reliable, plays a significant role in the information adoption [10, 54]. Indeed, unlike Ohanian's [34] findings, this study does not give support to the influence of all celebrity's credibility dimensions on the two dependent variables (information adoption and information dissemination) for both studies. Indeed, in the Hend Sabry's endorsement for Tunisia as a touristic destination to visit, only the perceived expertise of Hend Sabry has an effect on online information adoption. None of these celebrities' perceived credibility dimensions appeared to influence significantly the followers' information dissemination on Facebook.

Such a result seems to be a little bit surprising, but could be explained by some circumstantial factors. Tunisia destination was hit hard after the Arab spring revolution [55] and terroristic attacks may have elicited among people high-perceived risks. Thus, recommending the destination for others may look like "engaging the recommender's responsibility", even though vis-à-vis of Tunisian people. Investigating into such emotional and behavioral states accompanying the adoption and dissemination by SNS users of particular endorsed and shared information, would be very insightful for researchers on the effectiveness of cause-related endorsements, especially to find out some answers to interrogations such as: whom? For what purpose? When? In which context? Etc. Likewise, we think that celebrities themselves have to be careful when it comes to spontaneously or voluntary endorsing causes. Indeed they have to think about the degree to which they may be perceived as credible information's source and to which they may be influential on the potential target that might consider their action as a media stunt.

On the other hand, in Dorra Zarrouk's endorsement of the chocolate, only the celebrity's perceived trustworthiness was found to influence significantly the followers' information adoption and dissemination on Instagram. This result is consistent with previous studies' findings regarding the source's perceived trustworthiness impact on the receiver's information adoption [10]. Indeed they showed that the extent to which consumers perceive a source as reliable, plays a significant role in the information adoption.

It should be noted that some authors have already underlined the complexity to assess the impact of celebrity's credibility particularly in an online environment. In this regard, Metzger and Flanagin [14] contended that understanding credibility in the online environment is especially problematic as there are more than one possible "target" of credibility evaluation that often are at work simultaneously: the content, the information source, the site. For theses authors, the source and content of information are likely to interact in intricate ways to affect users' credibility judgments.

Despite the different purposes (cause-related versus product-related) and SNS platforms, our results show that the celebrity's perceived congruence has a positive effect on the consumer's adoption and dissemination of the endorsed information. These results are consistent with previous studies about the effect of offline or line celebrities' endorsements on brand beliefs and attitudes [33, 47, 56]. We consider such results of great relevance, as they put forward the necessity to take into account the fit celebrity/endorsed product/cause/service to increase the likelihood of achieving the endorsement objective.

Moreover, managers have to be cautious when using celebrity endorsement. they should primarily focus on the perceived congruence between the endorser and the endorsed product/service/cause. This means that the popularity or the notoriety of a celebrity, on its own, may be not sufficient to guarantee positive reactions among information receivers. A pre-assessment of his/her match-up with the product/brand, is then required. This prerequisite is even more critical within the SNS context as information dissemination (negative or positive) may grow up rapidly and may produce a negative effect, if the chosen endorser is not fitting.

Besides, a particular interest should be given by firms and brands to the community management function. Indeed, it helps to lay out a monitoring system of their e-reputation on SNS and consequently allows better accompanying of the celebrities in their spontaneous or paid endorsements.

The celebrities themselves should care about what they endorse. Indeed, as they are public figures, they have to be careful about not only the number of endorsed products/brands/causes but also their nature, which may not match up the image they want to impel of themselves or the image that people have of them. Having many endorsements might undermine the celebrity's credibility.

6 Conclusion, Research Limitations and Avenues

The study shows that celebrities' endorsements impact on the SNS users and their information adoption and dissemination depends mostly on the perceived congruence between the endorser and the object of the endorsement. As to the impact of the celebrity's perceived credibility, results show little empirical evidence on the followers' information adoption and dissemination. They also show that for Hend Sabry's endorsement, only the perceived expertise of this actress has an effect on the information adoption by her followers. For Dorra Zarrouk's endorsement, only her perceived trustworthiness significantly influences her followers' adoption and dissemination of the information about the chocolate she prefers. Clearly, more studies are needed to uncover the variables behind these unexpected results on SNS platforms (compared to traditional communication models). Future studies may take into account the followers individual factors such as: gender, age, cause/product involvement, perceived risk in adopting and disseminating endorsed object/information. Other studies addressing spontaneous/non-spontaneous endorsements are also required. In the present study, Dorra Zarrouk's perceived attractiveness was found to have a negative influence on her followers' information adoption. This result may be explained by the endorsed product, which is not necessarily fitting to the attractive character of this actress; specifically given that she endorses the wedding dresses and gowns of Elie Saab, the well-known Lebanese fashion designer.

Despite the interesting results, some limitations are identified. Some of these are related to the sample size, composition, and selection technique (convenience). Also, we used only female endorsers in this study and spontaneous endorsements (*a priory*), which might limit the results' generalizability. Indeed, the growing popularity of SNS and the explosion of celebrity's endorsements on these platforms, make it crucial to thoroughly investigate their impact on information processing and use.

References

1. McCracken, G.: Who is the celebrity endorser? Cultural foundations of the endorsement process. J. Consum. Res. **16**, 310–321 (1989)
2. Atkin, C., Block, M.: Effectiveness of celebrity endorsers. J. Advertising Res. **23**(1), 57–61 (1983)
3. Goldsmith, R.E., Lafferty, B.A., Newell, S.J.: The impact of corporate credibility and celebrity credibility on consumer reaction to advertisements and brands. J. Advertising **29**(3), 43–54 (2000)
4. Choi, S.M., Rifon, N.J.: Antecedents and consequences of web advertising credibility: a study of consumer response to banner ads. J. Interact. Advertising **3**(1), 12–24 (2002)
5. Mittelstaedt, J.D., Riesz, P.C., Burns, W.J.: Why are endorsements effective? Sorting among theories of product and endorser effects. J. Curr. Issues Res. Advertisement **22**(1), 55–65 (2000)
6. Ahmed, R.R., Seedani, S.K., Ahuja, M.K., Paryani, S.K.: Impact of celebrity endorsement on consumer buying behavior. J. Mark. Consum. Res. **16**, 12–20 (2015)
7. Wood, N.T., Burkhalter, J.N.: Tweet this, not that: a comparison between brand promotions in microblogging environments using celebrity and company-generated tweets. J. Mark. Commun. **20**(1–2), 129–146 (2014)
8. Erdogan, B.Z.: Celebrity endorsement: a literature review. J. Mark. Manage. **15**(4), 291–314 (1999)
9. Fleck, N., Korchia, M., Le Roy, I.: Celebrities in advertising: looking for congruence or likability? Psychol. Mark. **29**(9), 651–662 (2012)
10. Cheung, C.M., Lee, M.K., Rabjohn, N.: The impact of electronic word-of-mouth: the adoption of online opinions in online customer communities. Internet Res. **18**(3), 229–247 (2008)
11. Morris, M., Ogan, C.: The Internet as mass medium. J. Comput.-Mediated Commun. **1**(4), 39–50 (1996)
12. Leong, E.K.F., Huang, X., Stanners, P.-J.: Comparing the effectiveness of the website with traditional media. J. Advertising Res. **38**, 44–49 (1998)
13. Leeflang, P.S., Verhoef, P.C., Dahlström, P., Freundt, T.: Challenges and solutions for marketing in a digital era. Eur. Manag. J. **32**, 1–12 (2014)
14. Metzger, M.J., Flanagin, A.J.: Credibility and trust of information in online environments: the use of cognitive heuristics. J. Pragmat. **59**, 210–220 (2013)
15. Cho, C.H., Cheon, H.-J.: Why do people avoid advertising on the internet? J. Advertising **33**(4), 89–97 (2004)
16. van der Veen, R., Song, H.: Exploratory study of the measurement scales for the perceived image and advertising effectiveness of celebrity endorsers in a tourism context. J. Travel Tourism Mark. **27**(5), 460–473 (2010)

17. Sternthal, B., Phillips, L.W., Dholakia, R.: The persuasive effect of scarce credibility: a situational analysis. Public Opin. Q. **42**(3), 285–314 (1978)
18. Kaikati, J.G.: Celebrity advertising: a review and synthesis. Int. J. Advertising **6**(2), 93–105 (1987)
19. Friedman, H.H., Friedman, L.: Endorser effectiveness by product type. J. Advertising Res. **19**, 63–71 (1979)
20. Petty, R., Cacioppo, J., Schumann, D.: Central and peripheral routes to advertising effectiveness: the moderating role of involvement. J. Consum. Res. **10**, 135–146 (1983)
21. Agrawal, J., Kamakura, W.A.: The economic worth of celebrity endorsers: an event study analysis. J. Mark. **59**(3), 56–62 (1995)
22. Kamins, M.A.: Celebrity and noncelebrity advertising in a two-sided context. J. Advertising Res. **29**(3), 34–42 (1989)
23. Kamins, M.A., Gupta, K.: Congruence between spokesperson and product type: a matchup hypothesis perspective. Psychol. Mark. **11**(6), 569–586 (1994)
24. Stafford, M.R., Stafford, T.F., Day, E.: A contingency approach: the effects of spokesperson type and service type on service advertising perceptions. J. Advertising **31**(2), 17–35 (2002)
25. Ohanian, R.: The impact of celebrity spokespersons' perceived image on consumers' intention to purchase. J. Advertising Res. **31**(1), 46–54 (1991)
26. Bergkvist, L., Hjalmarson, H., Mägi, A.W.: A new model of how celebrity endorsements work: attitude toward the endorsement as a mediator of celebrity source and endorsement effects. Int. J. Advertising **35**(2), 171–184 (2016)
27. Hovland, C.I., Weiss, W.: The influence of source credibility on communication effectiveness. Public Opin. Q. **15**(4), 633–650 (1951)
28. Hovland, C., Janis, I.L., Kelley, H.H.: Communication and Persuasion. Yale University Press, New Haven (1953)
29. Kahle, L.R., Homer, P.M.: Physical attractiveness of the celebrity endorser: A social adaptation perspective. J. Consum. Res. **11**(4), 954–961 (1985)
30. Wen, C., Tan, B.C., Chang, K.T.T.: Advertising effectiveness on social network sites: an investigation of tie strength, endorser expertise and product type on consumer purchase intention. In: ICIS 2009 Proceedings, p. 151 (2009)
31. Keel, A., Nataraajan, R.: Celebrity endorsements and beyond: new avenues for celebrity branding. Psychol. Mark. **29**(9), 690–703 (2012)
32. Hoffman, S.J., Tan, C.: Following celebrities' medical advice: meta-narrative review. BMJ **347**, 1–6 (2013)
33. Wei, P.S., Lu, H.P.: An examination of the celebrity endorsements and online customer reviews influence female consumers' shopping behavior. Comput. Hum. Behav. **29**(1), 193–201 (2013)
34. Escalas, J.E., Bettman, J.R.: Connecting with celebrities: how consumers appropriate celebrity meanings for a sense of belonging. J. Advertising **46**, 1–12 (2017)
35. Ohanian, R.: Construction and validation of a scale to measure celebrity endorsers perceived expertise, trustworthiness, and attractiveness. J. Advertising **19**(3), 39–52 (1990)
36. O'Keefe, D.J.: How to handle opposing arguments in persuasive messages: a meta-analytic review of the effects of one-sided and two-sided messages. Ann. Int. Commun. Assoc. **22**(1), 209–249 (1999)
37. Knoll, J., Matthes, J.: The effectiveness of celebrity endorsements: a meta-analysis. J. Acad. Mark. Sci. **45**(1), 55–75 (2017)
38. McCormick, K.: Celebrity endorsements: Influence of a product-endorser match on Millennials attitudes and purchase intentions. J. Retail. Consum. Serv. **32**, 39–45 (2016)

39. Chung, S., Cho, H.: Fostering parasocial relationships with celebrities on social media: implications for celebrity endorsement. Psychol. Mark. **34**(4), 481–495 (2017)
40. Friel, C.: Celebrities finding new, lucrative ways to monetize their social network presence. Fox News, 19 August 2011
41. Breckler, S.J.: Empirical validation of affect, behavior, and cognition as distinct components of attitude. J. Pers. Soc. Psychol. **47**(6), 1191 (1984)
42. Sundar, S.S.: Exploring receivers' criteria for perception of print and online news. Journalism Mass Commun. Q. **76**(2), 373–386 (1999)
43. Chao, P., Wührer, G., Werani, T.: Celebrity and foreign brand name as moderators of country-of-origin effects. Int. J. Advertising **24**(2), 173–192 (2005)
44. Kanungo, R.N., Pang, S.: Effects of human models on perceived product quality. J. Appl. Psychol. **57**(2), 172 (1973)
45. Choi, S.M., Rifon, N.J.: It is a match: The impact of congruence between celebrity image and consumer ideal self on endorsement effectiveness. Psychol. Mark. **29**(9), 639–650 (2012)
46. Heckler, S.E., Childers, T.L.: The role of expectancy and relevancy in memory for verbal and visual information: What is incongruency? J. Consum. Res. **18**(4), 475–492 (1992)
47. Kamins, M.A.: An Investigation into the match-up hypothesis in celebrity advertising: when beauty may be only skin deep. J. Advertising **19**(1), 4–13 (1990)
48. Sussman, S.W., Siegal, W.S.: Informational influence in organizations: an integrated approach to knowledge adoption. Inf. Syst. Res. **14**(1), 47–65 (2003)
49. Fleck-Dousteyssier, N., Korchia, M., Louchez, S.: Les célébrités dans la publicité: le rôle de la congruence. Actes du 22 ème Congrès AFM **11**, 1–26 (2006)
50. Wu, C., Shaffer, D.R.: Susceptibility to persuasive appeals as a function of source credibility and prior experience with the attitude object. J. Pers. Soc. Psychol. **52**(4), 677 (1987)
51. Nunnally, Jum C.: Psychomtietric theory, 2nd edn. McGraw-Hill, New York (1978)
52. Kline, R.B.: Principles and Practice of Structural Equation Modeling. Guildford, New York (1998)
53. Burns, A.G., Bush, R.F.: Marketing Research, 3rd edn. Prentice Hall, Upper Saddle River (2000)
54. Petty, R.E., Cacioppo, J.T.: The elaboration likelihood model of persuasion. Communication and Persuasion, pp. 1–24. Springer, New York (1986)
55. Groizard, J.L., Ismael, M., Santana, M.: The economic consequences of political upheavals: the case of the Arab Spring and international tourism (2016)
56. Bruhn, M., Schoenmueller, V., Schäfer, D.B.: Are social media replacing traditional media in terms of brand equity creation? Manage. Res. Rev. **35**(9), 770–790 (2012)

What "Uses and Gratifications" Theory Can Tell Us About Using Professional Networking Sites (E.G. LinkedIn, Viadeo, Xing, SkilledAfricans, Plaxo...)

Karim Grissa[✉]

IAE-CEREGE Laboratory, University of Poitiers France, Poitiers, France
grissa.medkarim@yahoo.fr

Abstract. Social media continues to gain enormous popularity. Therefore, it is not surprising that this attracts the interest of researchers to study this cyber-social phenomenon particularly motivation issues as one of the most important axis in the social media literature. However, an analysis of literature concerning the motivations for using social media reveals a lack of studying professional networking sites compared to friendship-oriented social media (i.e. online social networks) and most particularly Facebook, despite the obvious interest that they present. Particularly as in the B2B context, interest in them exceeds that of online social networks (OSN) including even Facebook. Currently, a few number of researches who have become interested in studying the category of professional networking sites, but they did so from only the individual's point of view. However, it is also necessary to consider the motivations for using them from a company's point of view. To our knowledge, no research has been undertaken to study the motivations for using professional networking sites for both individuals and companies. For this purpose, we conduct a research by applying the 'Uses & Gratifications' Theory to explain the user behaviour in the field of professional networking sites. To collect data, 8 face-to-face interviews with companies and 5 focus groups with individuals were conducted.

The results permit to identify two types of motivations: contextual motivations and generic motivations in which a comparison with those previously found in the field of OSNs can be done.

Keywords: Uses & gratifications theory · Professional networking sites · LinkedIn · Viadeo · Uses companies motivations · Business networks

1 Introduction

In the last decade, social media has shown unprecedented growth and rapidly increased its popularity. This can be seen clearly from the rapid increase of their users and the advertising revenue generated. Facebook, Instagram, Myspace, Cywrold, HI5, Google+, Twitter, LinkedIn, Viadeo, and Xing are all examples of the success of what could be considered nowadays, as an unprecedented cyber social phenomenon. It is not surprising therefore that this success has sparked researchers' enthusiasm to study this phenomenon and that the fields of research into this phenomenon are many and varied. These research studies focus mainly on uses and behaviours [7], disclosure and private data [8], social interaction and

© Springer International Publishing AG 2017
R. Jallouli et al. (Eds.): ICDEc 2017, LNBIP 290, pp. 15–28, 2017.
DOI: 10.1007/978-3-319-62737-3_2

socialization [9] and the study of motivations [10–12]. This last research topic of motivations is without doubt one of the most important in social media literature and the reason that research specific to motivations for using social media has developed so quickly.

In fact, in the same way that research was previously made into the Internet context [13], and later into that of virtual communities [14, 15], by studying the motivations of usage, in their turn, investigations started into social networks. Thus, several researchers [1, 10] became interested in studying why Internet and mobile users use these web 2.0 platforms. However, these researches, focusing on the motivations for using social media, were made in general for all social networks. This, despite the fact that social media is characterized by a high intrinsic heterogeneity, thereby made their investigations together less consistent and irrelevant.

Consequently, this intrinsic heterogeneity of social media encouraged researchers [17, 18] to establish a cartography in order to present social media in homogeneous categories to facilitate their later studies. The most recent of these classifications is of particular interest because it is based on a major three-year study carried out by 28 researchers under the direction of the Mizuko Ito [19] (Digital youth project). This study reveals that for the goals of using social media are based essentially on two theories: Friendship and Interest (Fig. 1).

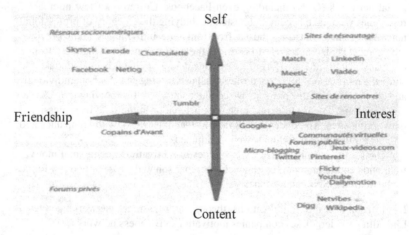

Fig. 1. Social Media Typology (Stenger and Coutant 2013)

In fact, this typology of social media highlights four categories, in which 2 typically are opposites: Friendship/sociability-oriented social media and Interest-oriented social media. However, analysis of the studies into the motivations for using social media reveals that they are largely dominated by the study of friendship-oriented social media (i.e. Online Social Networks) and more precisely Facebook for most of the research [1, 2]. Whilst the benefit of studying Facebook is undisputed professional networking sites in their turn are of no less importance, especially in B2B context:

- Huge audience size and high purchasing power: the audience for professional networking sites is also large (e.g. LinkedIn 470 million members…). In addition, the particularity of its audience is that it is composed of the most valuable target, as it belongs to a richer social class compared with other social networks (i.e. an audience that is older, more educated and belonging mostly to the professional world).

- Truth of the indicated digital identity: On virtual communities, members often use pseudonyms and on online social networks, members often prefer anonymity, however on professional networking sites a member's enrollment and participation is obliged to be done with their real offline identity.
- Lead conversion: professional networking sites clearly have the highest lead conversion rate in the B2B context. For this purpose, we can turn to the important study in 2015 made by HupSpot (international inbound marketing agency) which reveals that the conversion rate on LinkedIn is 2.74%, followed by Facebook 0.77% and Twitter 0.69%. This positions LinkedIn at the top of the list in terms of advertising efficiency in the B2B context, with a rate three times higher than that of Facebook and four times higher than that of Twitter.

Therefore, by considering the interest they present, the motivations determining the use of professional networking sites (i.e. Interest-oriented Social Media) deserve to be studied in the same way as it was the case for online social networks (i.e. Friendship-oriented Social Media). Particularly as, this interest exceeds that of the online social networks, including Facebook in the B2B and B2G context.

Thus, for all of the above reasons, professional networking sites (PNS) also deserve to be studied separately as it was the case for online social networks (OSN) particularly Facebook; especially as the many differences that characterize the two contexts (e.g. Goal/vocation of use, nature of published information, type of connected relationship, typical profile audience...) are likely to alter significantly the overall motivations hierarchy found in the context of ONS.

On the other hand, even the limited research studies that have been carried out recently on professional networking sites [4–6] have done so from only *the individual's point of view*. However, considering the particular interest of professional networking sites in the B2B context, it is also necessary to consider the study of the motivations of usage from a *company's point of view*. To our knowledge, no research has been carried out into the usage motivations of professional networking sites for both individuals and companies.

2 Literature Review

Since the exponential growth of social network sites, the focus of research into its motivations is among the most investigated in social media literature. Thus, [11, 12, 16] were among the first researchers to carry out an inter-social networks analysis[1] in order to investigate their motivations for use.

From this viewpoint, by mobilizing the theory of uses and gratifications (U&G), the research [12] analyzed the motivations for using the most popular local social networks in Norway (Underskog, Netby, Hamrungdan and Biip). They identified the following as the main motivations for using social networks: access to information, entertainment, social interaction and the management of one's personal identity. Always by using the

---

[1] In the sense that these researches include several social media at the same time in their samples.

same theoretical foundations (U&G), the research [11] demonstrates that researching information and making new friends are the most salient motivations for using Facebook and Myspace platforms. Similarly, concerning the analysis of research studies into inter-social networks, [20] studied the reasons why individuals refuse to use such platforms. He proposes that the major reasons for the decision not to subscribe or to unsubscribe an account are social grooming and gossip.

However, the contribution of these inter-social networks analysis research should be taken carefully because social networks are extremely diverse and heterogeneous. Thus, generalizing motivations to any type of social networks seems to be less significant and inconsistent. Because this can contain serious methodological risks, which probably would lead to some controversial results. Aware of such difficulties, some researchers have chosen to focus their research on studying just one social network. Consequently, online social networks belonging to the friendship-oriented social media category were preferred [10, 12] and in particular Facebook [2, 3, 13]. The domination of studies of Facebook within social media literature could be explained by various reasons: the huge audience (1,8 billion users), the possibility to make more precise commercial targeting, the high level of (in)voluntary disclosures of members and their data traceability, the high average member connection time, the massive presence of brands on this platform. These reasons to some extent explain why researchers focused particularly on Facebook.

To summarize, we note that the majority of researchers strongly agree that keeping up with existing relationships, making new friends, socializing and communicating with friends, searching for pleasure and leisure activities, passing time and avoiding boredom moreover searching and accessing to information, are the main motivations for using Facebook. Evidently, depending on the context and the nature of the research conducted, other more specific motivations are likely to appear, such as looking for romantic relationships [21] or contact surveillance or monitoring [22] and even research into promoting social events and festivities [3]. More recently a few researchers have begun to be interested in exploring which factors encourage the use of professional networking sites as an individual [4–6].

To summarize, Table 1 shows advances in research focusing on motivations in social media networks.

Table 1. Summarizing the literature review of using motivations

Internet	Virtual community	Social networks (Inter social networks analysis)	Online Social Networks (One social network analysis, mainly Facebook)	Professional networking sites
Papacharissi and Rubin (2000) Ferguson and Perse (2000) Kraut et al. (1998)	Ridings and Gefen (2004) Hiltz and Wellman (1997) Lamp, Velasquez and Ozkaya (2010)	Raacke and Bond-Raacke (2008) Brandtzoeg and Heim (2009) Tufecki (2008) Barker (2009)	Papacharissi and Mendelson (2011) Ellison, Steinfield and Lampe (2007) Dunne, Lowler and Rowley (2010) Special and Li-Barker (2012)	Grissa (2016) Florenthat (2015) Moeser (2013)

Finally, we note that different theories have been used to explain the user behaviour of these platforms from a socio-psychological perspective. The *'Uses and Gratification' theory* clearly appears to be the most utilized and also the most significant in social media literature to explain this behaviour.

2.1 Uses and Gratification Theory

The 'Uses and Gratifications' (U&G) theory is a socio-psychological approach which is epistemologically considered as positivistic. It assumes that individuals are active, rational and objective in their choices (in the sense of goal-oriented). As its name indicates, it seeks to answer the question of why these active and rational individuals use the various aspects and features of social media by the identification of gratification/rewards they seek. This is why historically the U&G theory was regularly used to explain the use of traditional media (i.e. Television, Magazine, radio, video games, Manga…). In the digital age, from the early 2000s, the use of this theory has been extended to explain the internet use phenomena [13], and then virtual communities [14, 15], and more recently social networks [1, 10, 11].

It is still worth noting however that the U&G theory is not exempt from criticism particularly in the context of traditional media [23]. These criticisms can be summarized in two broad categories:

- On the one hand it has a very optimistic and positive vision of media, owing to the belief that social media exists simply to satisfy the needs, desires and interests of people. Moreover, it also considers that individuals are completely free to select media that provides them with the most satisfaction. But the reality is sometimes totally the opposite to this simple assumption of the actor-audience. In fact, in some cases individuals found themselves forced to be followers and exposed to the power of media influence and its socio political content such as TV and Press.
- On the other hand, a very utilitarian perspective of individual behaviour concerning the choice of media. In fact, the assumption that media selection is simply the result of purely instrumental behaviour leads to the possibility of forgetting that sometimes the use of media could be the result of ritual behaviour. This could also lead to ignore the importance of the sociocultural environment and its ability to influence the media selection.

However, these limits remain relativized in the digital context because unlike traditional media, the Internet gives back more power to individuals. Thus, owing of its interactive features and its many tools, individuals have freer choice and become more active. This corresponds to one of the fundamental hypotheses on which the U&G theory is based. In addition, even the limit of its conceptualization of an individual characterized by too much instrumental behaviour (i.e. goal-oriented choice), becomes relativized because our context is professional networking sites, which are interest-oriented social media. That is why the use of U&G theory seems to be highly justified in helping us explaining the usage behaviour in the context of professional networking sites for both individuals and companies.

3   Methodology

This research aims to understand on the one hand, the motivations of usage adapted to professional networking sites (e.g. LinkedIn, Viadeo, Xing, Youpeek, SkilledAfricans...) and on the other hand, to understand these motivations for both individuals and companies (as the latter are still neglected in social media literature). Given the exploratory nature of our research objectives, a qualitative survey appeared to be the most appropriate tool. Thus, two qualitative data collection methods were conducted with companies and individuals:

- Companies: Eight face-to-face interviews were conducted, lasting between 30–45 min, mostly made in Paris, over a 5 months period, with company managers (e.g. a Digital Strategist, Social Media Managers, Senior Community Managers, a Digital Transformation Manager, Brand Managers). The object was to ask them about the motives and objectives of their company when creating and using an enterprise page on a professional networking site. To ensure the diversity of participating companies within our sample, 3 levels of enterprise page activity were chosen with the help of the Klout score measuring programme: 2 companies with a very active enterprise page use (Klout score > 70); 4 companies with a medium entreprise page use (Klout score [40, 60]) and 1 company with a light company page use (Klout score < 35). It should be noted that one company, not present on professional networking sites but very active on the online social networks (Facebook, Instagram, and especially Pinterest) was included in our sample group to gain a better understanding of the apparently reasons for not becoming a member of this social media category.
- Individuals: Five focus groups conducted in two specially equipped rooms properties of two academic research centers (the CEPE in Angoulême and the MSHS in Poitiers). Each focus group lasted between 1 h 30 min-2 h, and was constituted about 5 to 6 participants. The condition for participation in the study was simply to have a profile on a professional networking site. This condition was not satisfied for just one person in each focus group, in order to identify the perceived reasons that hamper the use of these professional networking sites. After completing our data collection, the contents were analyzed separately for each profile, by applying the thematic analysis method. Initially, the contents were fully transcribed. Then we extracted passages and after several readings, 'units of meaning' progressively began to be revealed. By using qualitative data analysis software (Nvivo 11), all the units of meaning were automatically associated with their verbatim. This coding work enabled themes to emerge that we carefully unified those that were sufficiently similar. This enabled us finally to identify patterns that corresponded to motivational factors. These motivational factors for both individuals and companies will be presented in a subsequent section.

4 Results

4.1 Company Motivations for Using Professional Networking Sites

Information diffusion and advertisement

The content analysis indicates that the use of enterprise page is motivated by the desire to create a brand community in order to apply afterward the marketing content of the company by creating as much possible commitment regarding the publications: "Our entreprise page has a dual goal: firstly, to build a community potentially interested in our services. Secondly, to animate it to increase the commitment to our content... Consequently this potential community becomes loyal to what we publish". For some companies running a business in the B2B context, it is sometimes their main channel for diffusing information: "It was a strategic choice for us if we really want to communicate effectively about our company, and particularly because we have a specific clientele that is difficult to reach via Facebook"

Increasing the e-reputation and online visibility of the company

The creation and use of a company page on this type of social network could be explained by the desire to provide a modern company image for the audience as well as optimizing the online visibility by the diversification of its social media presence. All of this, appears to be done within a coordinated strategy that intends to increase the e-reputation of the company."Our decision to use Viadeo is part of a global vision for our online company awareness. That is why, we are also present on Twitter, Facebook and we have also a chain on YouTube. This allows us to consolidate our efforts in SMO and additionally enhance the search engine position of our website and blog."

Doing business and prospecting its own market

Professional networking sites present many advantages to companies. In this context, companies indicate that these platforms presents a space that enables them to find partners as well as develop different ways of prospecting their market: "For us as a firm selling vigilance and surveillance equipment to companies and local communities, LinkedIn became our main tool for prospecting the potential market, besides our commercial team,...In fact, with LinkedIn Pulse, we can reach exactly the right customers we are looking for...."

Human resource solutions

Recruitment solutions are one of the motivations that particularly encourage a company to use this category of social media, whether with the aim of diversifying their recruitment channels: "One of the reasons why we created our company page is to be able to diffuse directly our internship and job offers", or with the intention of making it a main and unique recruitment channel for some technical jobs: "For some key jobs, we usually use the premium offer *Talent solutions* to find the best candidates and that works pretty well."

E-customer relation management and monitoring the market

The company use of professional networking sites is also explained by the desire to establish a relationship strategy with its customers in a B2B context, or simply by the need to be in the right place to control negative comments and moderate some of the positions taken against the brand. Furthermore, it makes it easier for a company to monitor its sector by checking competitors' pages and identifying the influential persons in their field of business activity (Table 2).

Table 2. Summarizing companies'motivations to use professional networking sites

Information diffusion and advertisement
Creating and animating a community around the brand
Communicating about the current news of the company and creating the buzz
Increase the e-reputation & online visibility of the company
Giving a modern image of the company
Improving the online visibility and its search engine optimization (SEO)
Doing business and prospecting its own market
Find partners and set up collaboration
Create another sales channel
Develop/expand the marketing actions
Human resource solutions
Recruit employees
Diversify the recruitment process
Being a strong employer branding
e-CRM and monitoring the market:
Optimize customer relationship experience and engaging him more
Moderate the shared negative comments and publications against the brand
Participate in groups relating to the business activity of the company
Surveillance the others enterprise competitor pages and their published contents
Identify and follow the influential experts in the firm business sector

4.2   Using Motivations of the Professional Networking Sites for Individuals

- Looking for a job/internship and performing its career management

This motivation quickly emerged in our content analysis. It may be manifested at different levels, whether in the search for a job/internship, or even in career management by searching for more interesting opportunities: "LinkedIn is very useful to me currently and even at the beginning of my professional life… By the way, it was thanks to LinkedIn that I found my internship. Now I like to be more 'on standby' to discuss other interesting experiences and new challenges by looking out for better opportunities for me as a software development engineer". Furthermore, they present an opportunity for members to promote around their professional identity.

- Building his personal branding and highlighting his skills

Professional networking sites are also used by individuals to confirm their expertise in their business activities through the continuous publication of content relating to their skills, as well as to try and optimize their online visibility: "when you are in a liberal profession such as a Freelancer, it is here that the business networks become more important to you than many others…because clients are not likely contact you before checking your expertise online… That is why you have to be active on this type of social network to achieve more self-exposure and expertise on the internet."

- Developing of his own professional network

Use of these social networks is also explained by the desire to develop one's own professional network composed mainly of classmates, working colleagues and people with whom the member shares the same skills. Thus, they are used to follow the recognized experts in one's business activity: "One of the reasons I was advised to focus on the use of Viadeo rather than LinkedIn is the facility to locate my professional network. As Viadeo is a local French business network, it was easier for me to find other students from my graduation year at university and also my work colleagues".

- Learning and exchange of information

Individuals also use these platforms to exchange views and debate with other professionals about issues of common interest. Thus, they have the opportunity to learn mutually from each other, besides the fact that they can benefit by following leaders in their field of specialization (e.g. the e-influencers programme on LinkedIn or webinars training courses) (Table 3).

Table 3. Summarizing Individuals' motivations to use professional networking sites

Looking for a job/internship and performing its career management
Searching for job/internship
Listening to opportunities (mission, freelance…)
Career development
Building personal branding and highlighting his skills
Confirming his expertise in the business sector
Enhancing his online visibility
Publish content related to his own skills
Developing of his own professional network
Find similarly profiles
Find old relationships (colleagues, students, classmates…)
Identify specialists/experts in his activity domain
Learning and exchange of information
Communicate and learn mutually from each other contacts
Participate in specializing groups related to his domain
Debate and exchange of their personal feedback experience
Follow the industry influencers

5 Discussion

Our research aims to study the motivations for using professional networking sites (i.e. interest-oriented social media), for both individuals and companies. The revealed results enable us to make some comparisons with those found in the field of online social networks. In fact, the results show two types of motivations:

- On the one hand, the generic/common motivations appear approximatively as the same in several contexts. Thus, they are resistible to the intrinsic heterogeneity of social media.
- On the other hand, the contextual motivations are entirely specific to the context being studied (i.e. professional networking sites).

Concerning the general common motivations, even though they are approximatively the same, when comparing the two contexts, their interpretation is likely to change relatively according to each context. For example, the information diffusion & advertisement; the improvement e-reputation & online visibility of the company are almost the same motivations of what an enterprise page on Facebook looks for. However, the way of achieving these motivations is not necessarily the same for each social network category. Certainly, unlike Facebook on which the tone of the diffused messages and content is rather more playful, pleasant and friendly, on Viadeo and LinkedIn the community manager in his editorial strategy will prefer a more serious and professional tone to disseminate his marketing content. It is the same, regarding frequency, periodicity and timing of diffusing marketing content in the professional networking sites context, which will be made with a light rhythm compared with Facebook, and more likely during working days (ideally at the beginning or end of the working day: 9–10 am or 4–5 pm). These differences also influence the features of the content marketing to adopt by the community manager in each type of social network, for example: the vividness of the content (e.g. video, photo, text…) and the degree of interactivity [24].

As much as it is for companies, the interpretation of the common motivations for individuals to use professional networking sites can vary partially according to the type of social network. For example, regarding development of one's own network motivation: on Facebook adding contacts to one's network is done between individuals belonging to the family circle and those in one's private life (friends, classmates…). While on LinkedIn the extension of the member's network concerns the professional world (work colleagues, people with whom one shares skills and professional issues…).

The same for exchange of information and learning motivation: For example, on online social networks (Facebook, Cyworld, HI5, Friendster…) members usually publish general or self-expressive information (e.g. date of birth, home city, hobbies, likes, holiday photos…) [3, 10] whereas on professional networking sites the published information is more about self-promotion.

On the other hand, the contextual motivations are totally different because they are specific and directly related to the study context. Thus, building one's personal e-branding and highlighting one's skills in addition to looking for a job or career development for individuals [3, 4], as well as recruitment and prospecting the market for companies are all contextual motivations because they are dependent on study context.

Concerning the influence of the importance of these motivations, the following figure may give a better explanation (Fig. 2):

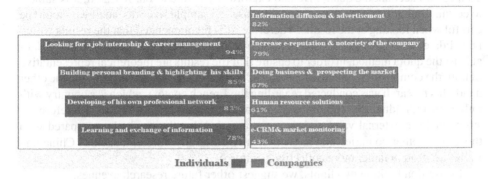

Fig. 2. Comparison between the motivations of Individuals members and Companies members

It is worth noting that all motivations display a significant effect for both individuals and companies. It remains however to note a lesser effect for e-CRM[2] and market monitoring motivations in using professional networking sites for companies. Nevertheless, all the motivations appear, in general, to be of purely instrumental/utilitarian nature, unlike the online social networks [1][3]. That is why for individuals, looking for a job and career development motivation is particularly salient [4]. However, this effect is reduced for companies, concerning human resources solutions motivation, at the level of being ranked in fourth position. This could be explained by the fact that recruitment and career development is a constant concern for individuals, whereas for companies this usually constitute a punctual and limited action in time.

To conclude, the instrumental and utilitarian nature of usage motivations is high with professional networking sites unlike online social networks, particularly Facebook [1]. In this context, we remind ourselves of the cartography of [17] that demonstrates that professional networking sites are mainly focused on 'interest'. Such findings enable us to relativize the limits formulated to the "U&G" theory, mainly in the context of traditional media (TV, press…), and more precisely, regarding its conceptualization of a very goal oriented individual whilst ignoring the media ritualistic use. However, given that our study context deals with professional networking sites that are nothing else interest oriented media, it is normal that these plat-forms are mainly instrumental.

It is similar to the limit of its conceptualization of an individual totally free of choice. This limit also becomes relativized in our study context because the internet in general and social media in particular, owing to their interactive features and the many tools made available for an individual's use, give back more power and freedom to individuals to make their choices. For all of these reasons, the "U&G" theory is proving to be relevant in explaining the user behaviour on professional networking sites.

---

[2] e-Customer Relation Management.

[3] We remind that Papacharissi & Mendelson results (2011) demonstrate that the use of the Facebook is mainly ritualistic.

6 Limits and Future Directions

Taking into account the qualitative and exploratory nature of our research, it is subject to certain limitations. On the one hand, because the sample size was small we should be careful when looking at the results. Therefore, it is recommended that the results would be validated through a larger scale quantitative sample using a probabilistic method, or at least the quota method, in order to ensure that the results are theoretical representative of both the composition of individuals and of companies in these platforms. On the other hand, this research was conducted regarding the French context, which is a country with rather an individualistic cultural dimension in the sense of Hofstede. Therefore, to strengthen the external validity, it is recommended that the results be compared with those from other countries that are sufficiently culturally different, such as China and India (i.e. from a rather more collectivist culture).

In addition to these two limits, we suggest other future research avenues:

Firstly, a quantitative survey appears to be the logical continuation of this research that would permit to determine empirically the most salient motivations of usage behaviour of professional networking sites besides to verify in what way some variables[4] are likely to modify the general hierarchy of these motivations. For example, by introducing the variable of the nature hierarchical position within the company, it would be legitimate to think that if we are investigating rather human resources managers, it is likely that the recruitment motivation will be the most relevant at the place of information diffusion and advertisement motivation which will be rather privileged by digital marketers and community managers.

Secondly, after revealing the motivations for using professional networking sites for both companies and individuals, it would be rather interesting now to focus on studying the factors that would encourage a rather more active, continuous and regular use of these platforms. In this context, we remind ourselves that the average time of an individual's daily connection on LinkedIn remains significantly lower than on Facebook which is evaluated to 46 min (Source: Facebook Business).

Thirdly, based on the results of [15] demonstrating that motivations for using virtual communities are not stable but progress through time, it would be reasonable to think that it is the same for social networks. Thus, a longitudinal research over 4 to 5 years would be the most suitable to highlight the evolution of these motivations based on the life-cycle of the member on the platform (i.e. Student in his last year at university, graduated student, junior manager and if possible senior manager).

Fourthly, though this research has focused on the study of the usage motivations of a member from a psychological perspective, it is no less important to extend the study to consider a sociotechnical perspective (technological factors) by studying the impact of the incentive mechanisms (e.g. platform email invitation, automated sponsorship...) as well as the attractiveness, originality of web design in the desire to use these platforms, as an example by studying the influence of interactivity, clarity and responsive design in the member intention to test these platforms or continue using them actively.

---

[4] e.g. for individuals: gender, professional status.../ for companies: company size, nature of the hierarchical position....

Acknowledgement. The author gratefully thanks his doctoral supervisors respectively Inés De La Ville and André Leroux.

References

1. Papacharissi, Z., Mendelson, A.: Toward a new(er) sociability: uses, gratifications and social capital on facebook. In: Stelios Papathanassopoulos (Ed.) Media Perspectives for the 21st Century (2011)
2. Joinson, N.: 'Looking at', 'Looking up' or 'Keeping up with' People? Motives and uses of facebook. In: Proceedings of the 26th Annual SIGCHI Conference on Human Factors in Computing Systems, pp. 1027–1036 (2008)
3. Special, W.P., Li-Barber, K.T.: Self-disclosure and student satisfaction with Facebook. Comput. Hum. Behav. **28**(2), 624–630 (2012)
4. Grissa, K.: What makes opinion leaders share brand content on professional networking sites (e.g LinkedIn, Viadeo, Xing, SkilledAfricans…). In: 2016 International Conference on Digital Economy (ICDEc), Carthage, pp. 8–15 (2016)
5. Florenthal, B.: Applying uses and gratifications theory to students' LinkedIn usage. Young Consumers **16**(1), 17–35 (2015)
6. Moeser, G., Morryson, H., Shwenk, G.: Determinants of online social business network usage behavior: applying the technology acceptance model and its extensions. Psychology **4**(4), 433–437 (2013)
7. Lampe, C., Ellison, N., Steinfield, C.: A Face(book) in the crowd: social searching vs. social browsing. In: Proceedings of the 2006 20th Anniversary Conference on Computer Supported Cooperative Work, pp. 167–170. ACM Press, New York (2006)
8. Acquisti, A., Gross, R.: Imagined communities: awareness, information sharing and privacy on the facebook. In: Proceedings of Privacy Enhancing Technologies Workshop, pp. 36–58, Cambridge (2006)
9. Steinfield, C., Ellison, N., Lamp, C.: Social capital, self-esteem, self-esteem, and use of online social network sites: a longitudinal analysis. J. Appl. Dev. Psychol. **29**(6), 434–445 (2008)
10. Dunne, A., Lowler, M.A., Rowley, J.: Young people's use of online social networking sites: a uses and gratifications perspective. J. Res. Interact. Mark. **4**(1), 46–58 (2010)
11. Raacke, J., Bond-Raacke, J.: MySpace and facebook: applying the uses and gratifications theory to exploring friend-networking sites. Cyberpsychol. Behav. **11**, 169–174 (2008)
12. Brandtzaeg, P.B., Heim, J.: Why people use social networking sites. In: Online Communities and Social Computing, pp. 143–152 (2009)
13. Papacharissi, Z., Rubin, A.M.: Predictors of internet use. J. Broadcast. Electron. Media **44**(2), 175–196 (2000)
14. Ridings, C.M., Gefen, D.: Virtual community attraction: why people hang out online. J. Comput. Medicated Commun. **10**(1) (2004)
15. Lampe, C., Wash, R., Velasquez, A., Ozkaya, E.: Motivations to participe in online community. In: 28th Conference on Human Factors in Computing Systems, pp. 1927–1936 (2010)
16. Barker, V.: Older adolescents' motivations for social network site use: the influence of gender, group identity, and collective self-esteem. CyberPsychol. Behav. **12**, 209–213 (2009)
17. Stenger, T., Coutant, A.: Médias sociaux: clarification et cartographie- pour une approche sociotechnique. Décisions Mark. **70**(24), 107–117 (2013)
18. Kaplan, A.M., Haenlein, M.: Users of the world, unite! The challenges and opportunities of social media. Bus. Horiz. **53**(1), 59–68 (2010)

19. Ito, M.: Hanging Out, Messing Around, and Geeking Out: Kids Living and Learning with New Media. The MIT Press, Cambridge (2010)
20. Tufekci, Z.: Grooming gossip, facebook and myspace. Inf. Commun. Soc. **11**(4), 544–564 (2008)
21. Tosun, L.P.: Motives for facebook use and expressing "true self" on the internet. Comput. Hum. Behav. **28**, 1510–1517 (2012)
22. Sheldon, P.: The relationship between unwillingness-to-communicate and students' facebook use. J. Media Psychol. **20**(2), 67–75 (2008)
23. Ruggiero, T.E.: Uses and gratifications theory in the 21st century. Mass Commun. Soc. **3**(1), 3–37 (2000)
24. Cvijikj, I.P., Michahelles, F.: Online engagement factors on facebook brand pages. Social Network Analysis and Mining (2013)

Intention of Adoption of Mobile Commerce from Consumer Perspective

Hela Ben Abdennebi[⊠] and Mohsen Debabi

ESCT, University of Mannouba, Tunis, Tunisia
h.b.abdennebi@hotmail.com, Mohsen_Debabi@yahoo.fr

Abstract. Mobile commerce (M-commerce) as an extension of e-commerce is a huge success in many countries. The number of mobile phone subscribers and revenues that are generated are growing rapidly, demonstrating the great potential of M-commerce. This may explain the eager of theorists to understand and explain the factors that explain the intention of its adoption. However, we noticed that many studies have been conducted using traditional adoptions theories and models that focus mainly on technological aspects and they ignored the fact that the user of m-commerce is a consumer. In this study we tried to explore the factors influencing the intention to adopt mobile commerce in Tunisia from value perspective. The results of this study show that perceived emotional value and perceived quality value are influencing the intention of adoption of mobile commerce in Tunisia. Several limitations and future directions of research are discussed.

Keywords: Mobile commerce · Consumer · Perceived value · Intention of adoption

1 Introduction

Mobile communication technologies have major impact on day-to-day life in the world (Puspitasari and Ishii 2016). The wide expansion of mobile devices in the world is a global phenomenon bringing significant economic consequences. Mobile technology belongs today to a new information technology generation that integrates very promising features. The features continue to improve and their prices to drop, making their acquisition increasingly easy, especially in a country like Tunisia. Indeed, developing Internet and mobile technologies provide innumerable service innovations for consumers; Laukkanen (2016) and created new opportunities to companies to interact and communicate with them. Mobile technology is then presented as a new commercial phenomenon that has given rise to a new generation of electronic commerce called mobile commerce (Zhang et al. 2012). However, Kim et al. (2007) explain that the main reason for the rapid growth of M-commerce is the rapid adoption of Mobile Internet as a medium of communication, contents service and commerce. In addition, the growth of m-commerce is closely linked to that of m-Internet. Therefore, a clear and comprehensive understanding of m-Internet adoption is essential to study customer behavior related to m-commerce. Hence, in this study we will try to focus on mobile internet as a mobile commerce service.

© Springer International Publishing AG 2017
R. Jallouli et al. (Eds.): ICDEc 2017, LNBIP 290, pp. 29–40, 2017.
DOI: 10.1007/978-3-319-62737-3_3

The global success of mobile technology then incited practitioners to place it under their focus and more new mobile services with high added value are flourishing. As for theorists, they tried to develop research models that explain the adoption behavior of this new technology.

To explain mobile commerce adoption, several authors made recourse to traditional theories such as the technology adoption model (TAM) (Davis 1989), planned behavior theory TPB) (Ajzen 1991) or Diffusion of Innovation (DOI) theory (Rogers 1983). Indeed, several studies examining mobile technology, in particular services, used an extension of different theories to explain intention to adopt different technologies. However, several authors like Nysveen and Pedersen (2002), Pedersen and Ling (2002), Yu and Liu (2003), Kim et al. (2006), Nysveen et al. (2005b) indicated that traditional models cannot fully explain the factors that affect intentions to adopt or reject mobile commerce services. The Main reason for this failure because the mobile commerce user is mainly a consumer and is not just a technology user. (He/she pays for the service in question. He/she uses the service voluntarily and not necessary for professional purposes so the service is not free or provided by their companies). Then, forgetting this fact can give an inaccurate assessment of mobile commerce adoption (Al Hinai et al. 2007).

To understand Mobile commerce adoption from consumer perspective, some studies used perceived value (Kim and Han 2011; Wang et al. 2013; Yang and Jolly 2009; Wang et al. 2013). However, Kleijnen et al. (2007) state that mobile services providers do not understand how consumers perceive value of mobile technology. Specifically, it is unclear how value is constructed by the consumer (Sandstrom et al. 2008). This could be explained by the fact that the literature of perceived value of mobile technology is very scarce (Spink et al. 2011). Moreover, Laukkanen (2016) explain that to date, little research examines the factors inhibiting the adoption process behavior toward innovation. Hence, the present study tries to bring some answers to understand the intention of adoption of mobile commerce from value perspective.

2 Theoretical Framework

Kim et al. (2007) has developed a value–based adoption model VAM of m-internet. The crux of VAM is the value construct, which is postulated to predict intention of adoption of m-internet. The result of their study was consistent with prior research on perceived value which has recurrently verified perceived value as a predictor of intention. Hence, to understand the intention of adoption of mobile internet in Tunisia, from consumer perspective, we will try to understand the perceived value as antecedent of intention adoption.

3 Intention of Adoption

Since the studies of Fishbein and Ajzen (1975), Davis (1989), the concept of intention continues to attract the attention of researchers in Marketing and Information Systems. Moreover, social psychology research suggests that intention would be the best predictor of behavior because it allows researchers to probe individually all the important factors that could influence individuals' behavior (Fishbein and Ajzen 1975). According to TRA and TPB, intention is central to studying behavior. Similarly, Ajzen (1991) states that "Behavioral intentions are supposed to capture the motivational factors that influence behavior, they are indicators of the intensity of the will to try, of the effort one is willing to make to behave in a certain way". Moreover, Fishbein and Ajzen (1975) argue that intention is a desire, a wish, a determination or a willingness to behave. Intention then allows for predicting behavior. The stronger it is, the higher is the likelihood to induce an effective behavior. Meanwhile, Triandis (1979) believes that "intentions respond to instructions given by the individual to behave in a certain way". In summary, intention seems to be an appropriate factor predicting information systems adoption behavior in general and could be used to predict mobile technology adoption.

3.1   Perceived Value

Perceived value is an important construct for understanding consumer responses to existing and emerging mobile services and it is one of the most important determinants of behavioral intentions to use the technology (Kuo and Yen 2009).

However, no definition of perceived value really enjoyed unanimity of the research community to date (Amraoui 2005). In a review, Woodruff (1997) in Gharbi and Souissi (2003) defines perceived value as "a preference and an assessment made by the customer, of product (or experience) attributes, its performance and the outcomes of its use (or experience), facilitating or hindering the goals and the outcomes that the individual wants to achieve in practical situations".

Several distinctions of perceived value have been proposed in the literature. Tauber (1972) argues that perceived value of a purchase process can be utilitarian, hedonistic, social or psychological. Burns (1993) identified four types of value. These are product value, use value, possession value and overall value. Babin et al. (1994) distinguish between utility value and hedonic value of shopping experiences. Finally, Sheth et al. (1991) identified five categories of perceived value, i.e. epistemic, functional, social, emotional and conditional. Sweeney and Soutar (2001) developed a scale to measure the last four types of perceived value.

As for its dimensions, the concept of perceived value has evolved from a bi-dimensional to a multidimensional construct. Indeed according to Zeithaml (1988), the bi-dimensional conception of perceived value refers to an evaluation made by the consumer between what they receive (the positive dimension of value) and what they give in return (the negative dimension of perceived value).

Regarding the three-dimensional approach, Amraoui (2005), following Sweeney and Soutar (2001), distinguishes three main dimensions of perceived value: economic value, social value and emotional value. From these three dimensions of value, this author developed a measurement instrument of different durable products. Amraoui (2005) found that emotional value is a unique dimension of perceived value (it is in fact influenced by the other two dimensions of value) and is the best indicator of consumer purchasing intent.

In addition, Sweeney and Soutar (2001) consider four dimensions of perceived value namely: emotional value, social value, functional or quality value and price or monetary value.

In what follows, we present each of the dimensions of perceived value.

Emotional value: Value derived from sentimental and emotional states that a product may procure.

Social value: Value derived from the product's ability to increase its own social recognition.

Price or monetary value: Value derived from short term and long term cost reduction.

Functional value: value related to the performance/ quality that fits the value that comes from perceived quality and expected performance of the product.

We will retain the concept of perceived value proposed by Sweeney and Soutar (2001) as adapted by Lee et al. (2002) to study mobile Internet.

Examining E-commerce, Gharbi and Souissi (2003) showed that consumers place value on their individual gains in the short-term. These values are emotional, social, and relate to quality and price.

M-Internet users may feel fun and excitement by trying a new technology or a new application. Studying mobile services, Nysveen et al. (2005a) found that perceived pleasure acts positively on intention. Studying mobile Internet, Batti and Ammami (2007) argue that emotional perceived value seems to positively affect the intention to adopt mobile Internet for most respondents. The author argues that evoking mobile Internet arouses in the respondents' joyful, pleasure and entertainment sensations. Hence the following hypothesis:

H1.1. Perceived emotional value has a direct positive impact on the intention to adopt mobile Internet.

Batti and Ammami (2007) state that perceived social value determines the intention to using mobile Internet to the extent that it allows them to improve their social image. In addition, mobile Internet allows some to strengthen ties with their relatives and colleagues.

H1.2. Perceived social value has a direct positive impact on the intention to adopt mobile Internet.

As for the perceived price value, Lee et al. (2002) explain that monetary value should not be neglected when studying mobile Internet since users have to pay relatively expensively to surf the net while stationary or traditional Internet services can often be accessed free of charge especially in the workplace.

H1.3. Perceived price value has a positive impact on the intention to adopt mobile Internet.

For perceived quality value, according to Lee et al. (2002) this concept is very important, and associates the functional value with a practical or a technical return that users can obtain by using mobile Internet. In other words, individuals use mobile Internet because of its useful functions.

For Batti and Ammami (2007), perceived quality or functional value seems to be a determinant of the intention of using the mobile Internet. He adds that most respondents combine high functional value to the mobile Internet because it allows them to access their e-mail box at any time and at any location, especially for those who do not have a traditional Internet connection.

Accordingly, we propose the following hypothesis:

H1.4. Perceived quality value has a positive impact on the intention to adopt mobile Internet.

4 Methodology

Data were collected by means of administering a questionnaire to a sample of 300 people. We opted for the convenience sampling method. The sample consists of 66.3% students, 31, 3% employees and the remaining 2.3% job hunters. Respondents' age spans from 18 to 35 years. Of the respondents, 61.7% are females and 38.3% are males. Most respondents have a mobile phone and access mobile Internet (See Appendix A).

To measure intention, we used the scale of Shin (2007) originally developed by Davis (1989). This scale was adapted to the context of mobile Internet by this author. This scale is one-dimensional and consists of four items. It has good reliability (Cronbach's alpha above 0.8) and validity.

To measure perceived value, we used the scale of Gharbi and Souissi (2003) adapted from Sweeney and Soutar (2001). The choice of this scale is justified by the fact that it takes into account the multidimensional nature of perceived value on the one hand and it has a good validity on the other (Gharbi and Souissi 2003). Moreover, this very scale was used to study mobile Internet by Lee et al. (2002) and has a very good reliability (alpha equal to 0.939). This scale consists of 15 items measuring the four dimensions of perceived value, i.e. perceived quality value, perceived social value, perceived emotional value and finally perceived price value. All our measurement scales (see Appendix A) were selected, translated and adapted to mobile Internet. The scale, in large part, uses a 5-point Likert scale ranging from 1 - "Strongly disagree" to 5 - "Strongly agree".

5 Analysis of Results

In this study, we used structural equation methods under the Partial Least Squares approach (PLS) using the SmartPLS 2.3 software.

Using the PLS method led us to estimate our measurement and the structural models. In what follows, we estimate the quality of these two models.

Estimating the measurement model: According to Fernandes (2012), the measurement model, also called external or "outer model" is assessed using the following criteria: reliability, internal consistency, convergent validity of the measures of the constructs and discriminant validity.

For reliability, Cronbach's alpha and Composite Reliability (CR) are greater than 0.7. In our study all these values are higher than 0.7 which shows a good internal consistency (See Appendix A). For validity, it amounts to estimate the discriminant and convergent validity of the constructs. For discriminant validity, the square roots of the AVE are greater than the correlations of the other latent variables. This represents good discriminant validity. Convergent construct validity is estimated by ensuring that AVE exceeds the threshold of 0.5 (Fornell and Larker 1981). (see Appendix A).

The structural model, also called internal or "inner model" is assessed to check for the predictive relevance of the latent variables, i.e. their nomological validity. This amounts to estimating Multiple R^2 and a Goodness-of-fit index, a comprehensive estimation index of PLS (Fernandes 2012). For this author and to ensure the fit quality of the structural model, we should examine the values of explained variance R2 and the overall GoF index (Goodness of fit).

The coefficients of determination (R2) account for the explained variance of the endogenous variables (SmartPLs gives a value of zero for exogenous variables).

The thresholds established by Wetzels et al. (2009) (low = 0.02;. Medium = 0.13 and high = 0.26). Chin (1998) considers that the standardized structural coefficients should be at least equal to 0.20 and ideally above 0.3 to be considered significant.

The obtained R2 values (See Appendix A) all exceed the threshold of 0.20. They should then confirm that the measured construct helped to explain a sizable share of explained variance. PLS provides a single fit index, the GoF index, which is the geometric mean of average communality and average R2, it is between 0 and 1. The GoF index for our structural model is 0.481 (greater than 0.36), representing a very good fit according to Wetzels et al. (2009).

6 Interpretation of the Results

Estimating the structural model through a Bootstrap procedure allows us to test our research hypotheses on the direct effects of the independent variables on the dependent variable (adoption of mobile Internet). The results will be discussed in what follows.

To confirm the direct effect of an independent variable on a dependent variable, we examine the regression coefficients estimating the so-called β Path coefficient. The direction of the relationship is given by the sign of the estimated parameter (β).

A relationship is said to be significant if the "t-Student" test associated with a standardized multiple regression coefficient is significant. SmartPLS software does not give the p-values of the t-Student test. Then, we compared the t-values to thresholds reported in Student table.

- A 5% error threshold (t-value = 1.96).
- A 10% error threshold (t-value = 1.64).

6.1 Impact of Perceived Value on the Intention to Adopt

We found that perceived price value has no impact on mobile internet adoption intent. This result was found by Hsiao (2013) for mobile internet. Al-Debei and Al-Lozi (2014) explain that people will be willing to use the mobile service only if the associated costs are reasonable because they are usually used for personal needs and thus the cost is borne mainly by them. In Tunisia, mobile internet price is very expensive so Tunisian are less likely to perceive price value, especially in the current modest economic situation of the country.

Perceived emotional value has a positive impact on intention. This finding is consistent with the results Ozturk et al. (2016) for mobile service (MHB). Therefore, the user could have felt emotions of joy and excitement with Internet access while using their mobile devices.

Perceived quality value has a positive impact on intention. This finding is similar to that found by Spink et al. (2011) while studying mobile services. With the introduction of 3G technology and the 4G technology users could perceive the functional usefulness of mobile Internet.

Perceived social value has no impact on the intention to adopt mobile Internet. This is consistent with the findings of Wang et al. (2013) and Batti and Ammami (2007) studying mobile Internet adoption in the Tunisian context.

7 Discussion and Conclusion

The results of this study revealed that emotional and quality value dimensions are the best predictors of mobile commerce adoption in Tunisia. Our findings have several important implications; marketing managers must pay much attention to the hedonic and functional aspects of mobile commerce. The user attaches greater importance to emotions and feelings that can be aroused while using Internet and the user attaches less importance to connection quality, speed, usability, etc. Marketing managers need to enhance users' positive feelings (e.g. pleasure and fun) so they can attract more users to adopt mobile commerce. Moreover, Mobile Internet providers are expected to focus on mobile Internet usability by highlighting the various uses of m-Internet. This might help employees to perform their work (sending urgent mails, accessing their mail boxes, etc.). For students, mobile Internet is the best tool for them to look for information, anywhere and anytime and stay connected to social networks and follow news, etc. In addition, service internet providers need to revise the price of mobile internet because consumers in general are usually concerned about the prices of such services (Kim et al. 2007) especially in Tunisia with economic challenges and many Tunisians face unemployment, poverty, etc.

There are several implications from this research. Firstly, this study comes to enhance our understanding of the factors that explain the intention of adoption of mobile commerce in Tunisia from consumer perspective and we tried to join previous calls for more empirical tests in the m-commerce area. Secondly, few studies were conducted to understand the effect of perceived value dimensions on the intention of adoption of mobile commerce within the context of developed countries but within the context of developing countries, they were scarce. Thus, this study tend to contribute in this domain.

Despite its potential contribution, our study has some limitations. In fact, we did not take into account some variables while constructing the model, like satisfaction and trust. These two variables could have been key variables in this study. Another limitation is the use of a convenience sample in the final data collection. Using a probabilistic sampling method could certainly improve generalizability of our results. However, we preferred a convenience sample because of accessibility and the ease in collecting data.

Many future research avenues are possible to identify, as mobile commerce is a relatively unexplored area. Indeed, research on intention to adopt m-commerce is scarce. Therefore, it would be interesting to conduct other studies especially that our present study does not claim to provide a definite and complete answer to the studied issue. For example, it is recommended to focus on a specific category of respondents. According to several authors, teenagers are the largest group that uses mobile commerce. Indeed, mobile professionals believe that teenagers are the rescuers of mobile commerce. They represent a changing dynamics and they are increasingly inclined to buy products and services through their mobile devices. One should pay attention to this segment users. In addition, at some future time, mobile commerce will still develop further, and it would be better to choose another mobile commerce feature other than mobile Internet to understand the intention of its adoption. We can of mention mobile games, mobile TV, etc.

A. Appendix

(See Tables 1, 2, 3, 4 and 5).

Table 1. Sample characteristics (N = 300)

		No	%
Use of internet	Yes	243	81
	No	57	19
Gender	Male	115	38.3
	Female	185	61.7
Current profession	Student	199	66,3
	Employee	94	31,3
	Unemployed	7	2,3
Age	18–20	22	7,3
	21–25	164	54,7
	26–30	97	32,3
	>35	17	5,7
Education	High School	19	6.3
	Bachelor degree	171	57
	Master, PhD	110	36.7
Total		300	100%

Table 2. Reliability and convergent validity of the measurement scales

	AVE	Composite Reliability	R Square	Cronbach's Alpha	Communality	Redundancy
Intention to adopt	0,650882	0,881672	0,415991	0,820954	0,650882	0,117094
Perceived price value	0,729664	0,889049		0,815031	0,729664	
Perceived quality value	0,639615	0,898603		0,858857	0,639615	
Perceived social value	0,830530	0,951460		0,932315	0,830530	
Perceived emotional value	0,606742	0,858867		0,790335	0,606742	

Table 3. Correlation between the latent variables to show the discriminant validity of constructs (on the diagonal are the square roots of the AVE of constructs.

	Intention	PPV	PQV	PSV	PEV
Intention	**0,8091**				
PPV	0,1536	**0,8445**			
PQV	0,5063	0,4878	**0,8265**		
PSV	0,2293	0,3988	0,5585	**0,9352**	
PEV	0,5645	0,3530	0,6051	0,5124	**0,86897**

Table 4. Scales and Items

Scales	Items
1. Intention to adopt	1.1. I intend to use mobile internet n
	1.2. I recommend people to use Mobile Internet
	1. 3. I intend to use Mobile Internet as soon as possible
	1. 4. I intend to carry on using Mobile Internet in the future
2. Perceived social value	2.1. The fact of using mobile internet allows me to give a good impression to my entourage
	2.2. The use of mobile internet improves the way am socially perceived.
	2.3. The fact of using mobile internet allows me to be socially proud
	2.4. The fact of using mobile internet helps me to feel socially acceptable
3. Perceived emotional value	3.1. I entertain myself with mobile internet
	3.2. I feel relaxed when I use mobile internet.
	3.3. Mobile internet makes me feel good
	3.4. Mobile internet gives me pleasure
4. Perceived price value	4.1. Mobile internet offers a good quality/price ratio
	4.2. Mobile internet may be accessed with reasonable prices
	4.3. Mobile internet offers much value for its access price
5. Perceived quality value	5.1. Mobile internet service is consistent
	5.2. Mobile internet service is interesting
	5.3. Mobile internet service has an acceptable high quality
	5.4. Mobile Internet meets my needs
	5.5. Mobile internet service has a consistent performance

Table 5. Results of the impact of the independent variables on the intention to adopt

Hypotheses		β	t-	Acceptance/rejection of hypothesis
H1.1	PEV-IN	0,310	2,087	Accepted
H1.2	PSV-IN	−0,096	0,605	Rejected
H1.3	PPV-IN	−0,130	1,020	Rejected
H1.4	PQV-IN	0,290	1,828	Accepted

References

Ajzen, I.: The theory of planned behavior. Organ. Behav. Hum. Decis. Process. **50**(2), 179–211 (1991)

Al-Debei, M.M., Al-Loz, E.: Explaining and predicting the adoption intention of mobile data services: a value-based approach. Comput. Hum. Behav. **35**, 326–338 (2014)

AlHinai, Y.S., Kurnia, S., Johnston, R.B.: A literature analysis on the adoption of mobile commerce services by individuals. In: Proceedings of the 13th Asia Pacific Management Conference, Melbourne, Australia, pp. 222–230 (2007)

Amraoui, L.: Les effets du prix, de l'image du point de vente et du capital de marque sur la valeur perçue des produits, Thèse de doctorat en Sciences de Gestion, IAE de Toulouse (2005)

Babin, B.J., Darden, W.R., Griffin, M.: Work and/or fun: measuring hedonic and utilitarian shopping value. J. Consum. Res. **20**, 644–656 (1994)

Batti, M., Amami, M.: Les facteurs influençant l'adoption du mobile commerce, Colloque AIM (2007). http://www.aim.asso.fr/index.php/mediatheque/view.download/339

Chin, W.: The partial least squares approach to structural equation modeling. In: Marcoulides, G.A. (ed.) Modern Methods for Business Research, pp. 295–336. Lawrence Erlbaum Associates, Mahwah (1998)

Davis, F.D.: Perceived usefulness, perceived ease of use, and user acceptance of information technology. MIS Q. **13**(3), 319–340 (1989)

Fernandes, V.: En quoi l'approche PLS est-elle une méthode a (re)-découvrir pour les chercheurs en management? Management **15**(1), 101–112 (2012)

Fishbein, M., Ajzen, I.: Belief, attitude, intention and behavior: an introduction to theory and research. Addison-Wesley Publishing Company, Reading (1975)

Fornell, C., Larker, D.: Evaluating structural equation models with unobservable variable and measurement error. J. Mark. Res. **18**(1981), 39–50 (1981)

Gharbi, J.E., Ben Souissi, S.: Quelle stratégie de réduction de l'impact du risque perçu de l'achat par Internet? Les quatrièmes Journées Internationales de la recherche en sciences de gestion: Ethique(s), Incertitude(s) et Changement(s), Association tunisienne des Sciences de Gestion, Hammamet, 11–13 mars 2003

Hsiao, K.-L.: Android smartphone adoption and intention to pay for mobile internet: Perspectives from software, hardware, design, and value. Libr. Hi Tech **31**(2), 216–235 (2013)

Kim, M., Jee, K.: Characteristics of individuals influencing: adoption intentions for portable internet service. ETRI J. **28**(1), 67–76 (2006)

Kim, H.W., Chan, H.C., Gupta, S.: Value-based adoption of mobile internet: An empirical investigation. Decis. Support Syst. **43**(1), 111–126 (2007)

Kim, B., Han, I.: The role of utilitarian and hedonic values and their antecedents in a mobile data service environment. Expert Syst. Appl. **38**(2011), 2311–2318 (2011)

Kleijnen, M., de Ruyter, K., Wetzels, M.: An assessment of value creation in mobile service delivery and the moderating role of time consciousness. J. Retail. **83**(1), 33–46 (2007)

Kuo, Y.-F., Yen, S.-N.: Towards an understanding of the behavioral intention to use 3G mobile value-added services. Comput. Hum. Behav. **25**(2009), 103–110 (2009)

Laukkanen, T.: Consumer adoption versus rejection decisions in seemingly similar service innovations: the case of the Internet and mobile banking. J. Bus. Res. (2016). doi:10.1016/j. jbusres.2016.01.013

Lee, Y., Kim, J., Lee, I., Kim, H.: A cross-cultural study on the value structure of mobile internet usage: comparison between Korea and Japan. J. Electron. Commer. Res. **3**(4), 227–239 (2002)

Nysveen, H., Pedersen, P.E., Thorbjornsen, H.: Explaining intention to use mobile chat services: moderating effects of gender. J. Consum. Mark. **22**(5), 247–256 (2005b)

Nysveen, H., Pedersen, P.E.: The adoption of a mobile parking service: instrumentality and expressiveness. SNF-Working Paper no. 76/02. Foundation for Research in Economics and Business Administration, Bergen, Norway (2002)

Nysveen, H., Pedersen, P.E., Thorbjornsen, H.: Intentions to use mobile services: Antecedents and cross-service comparisons. J. Acad. Mark. Sci. **33**(3), 330–46 (2005a)

Ozturka, A., Nusairb, K., Okumusa, F., Rosen, N.H.: The role of utilitarian and hedonic values on users' continued usage intention in a mobile hotel booking environment. Int. J. Hospitality Manage. **57**, 106–115 (2016)

Pedersen, P., Ling, R.: Mobile end-user service adoption studies: a selective review, HiA, Grimstad (2002). http://ikt.hia.no/perep/pedersen_ling.pd

Rogers, E.M.: Diffusion of Innovations, 3rd edn. The Free Press, New York (1983)

Puspitasari, L., Ishi, K.: Digital divides and mobile Internet in Indonesia: impact of smartphones. Telematics Inform. **33**, 472–483 (2016)

Sandström, S., Edvardsson, B., Kristensson, P., Magmusson, P.: Value in use through service experience. Managing Serv. Qual. **18**(2), 112–126 (2008)

Sheth, J., Newman, B.I., Gross, B.L.: Why we buy what we buy: a theory of consumption values. J. Bus. Res. **22**(2), 159–170 (1991)

Shin, D.H.: User acceptance of mobile internet: implication for convergence technologie Interact. Comput. **19**, 472–483 (2007)

Spink, P., Rahim, M.M., Singh, U.B.: User intentions of downloading games on mobile phones: an empirical evaluation using consumption value model. In: ACIS 2011 Proceedings. Paper 32 (2011)

Sweeney, J.C., Soutar, G.N.: Consumer perceived value: the development of a multiple item scale. J. Retail. **77**, 203–220 (2001)

Tauber, E.M.: Why do people shop? J. Mark. **36**, 46–59 (1972)

Triandis, H.C.: Values, attitudes, and interpersonnal behavior. In: Page, M.M. (ed.) Nebraska Symposium on motivation: Beliefs, Attitudes, and values. University Nebraska Press (1979)

Wang, H-Y., Liao, C., Yang, L.-H.: What affects mobile application use? The roles of consumption values. Int. J. Mark. Stud. **5**(2) (2013) ISSN 1918-719X, E-ISSN 1918-7203

Wetzels, M., Odekerken-Schröder, G., Van Oppen, C.: Using PLS path modeling for assessing hierarchical construct models: Guidelines and empirical illustration. MIS Q. **33**(1), 177–195 (2009)

Yang, K., Jolly, L.D.: The effects of consumer perceived value and subjective norm on mobile data service adoption between American and Korean consumers. J. Retail. Consum. Serv. **16**(6), 502–508 (2009)

Yu, J.L., C.-S., Liu, C. (2003), «Technology acceptance model for wireless Internet», Internet Research,13(3): 206

Zeithaml, V.A.: Consumer perceptions of price, quality, and value: a means-end model and synthesis of evidence. J. Mark. **52**(2), 2–22 (1988)

Zhang, L., Zhu, J., Liu, Q.: A meta-analysis of mobile commerce adoption and the moderating effect of culture. Comput. Hum. Behav. **28**(2012), 1902–1911 (2012)

Not Always a Co-creation: Exploratory Study of Reasons, Emotions and Practices of the Value Co-destruction in Virtual Communities

Arij Jmour[(⊠)] and Imen Charfi Ben Hmida

Faculty of Economics and Management Science, Sfax, Tunisia
arijjmour@gmail.com, charfiimen@yahoo.fr

Abstract. This exploratory research proposes to examine reasons, emotions and practices of the value co-destruction in the virtual community. A qualitative exploratory study using the netnography method combined with the critical incidents technique was conducted. The results clarify, firstly, that the corporate misbehavior, the unfairness of the firm and the failure of the consumer experience are the main reasons of the value co-destruction, and secondly, that many emotions for example anger and feelings of betrayal can conduct to negative consumer practices. These practices can destroy the value in the virtual community. We mainly cite the boycott of the brand, the vengeance of Internet users, and the reduction of patronage in particular the comparison with other competing brands.

Keywords: Value co-destruction · Virtual community of the brand · Netnography · Causes · Emotions · Practices

1 Introduction

With the advent of the virtual brand communities, previous studies tend to focus on the positive results of the value co-creation [16] that some researchers [5, 12, 24] emphasized that the practices of this co-creation reinforce interaction between community members and understanding depth of the brand. However, research in the inverse possibility of the value co-creation in virtual brand community is understudied. We talk about the value co-destruction that is becoming an important mean to conceptualize the non-positive results of interactions between actors [16]. Indeed, some researchers [15, 17] investigating the value co-destruction from a service-dominant-logic perspective, have founded the fact that interactions between service systems can not only co-create the value, but also destroy it. However, the issue of the value co-destruction in the virtual community is largely ignored. By studying the value co-destruction, researchers have concentrated on misuse of resources [15, 18]. However, few studies focus on actors (supplier and consumer) and how they apply to negative interactions [16]. Although recent studies have focused on co-creating the value and valuing experiences through technology of information and communication, there is less knowledge about the value co-destruction when technology of information and communication are used [13]. The role of technology has

© Springer International Publishing AG 2017
R. Jallouli et al. (Eds.): ICDEc 2017, LNBIP 290, pp. 41–54, 2017.
DOI: 10.1007/978-3-319-62737-3_4

not been considered in the value co-destruction to date. In addition, Algesheimer et al. [1] found that the community's positive influences give birth to its negative influences and that these communities can negatively influence their members. Laroche et al. [11] and Adjei et al. [2] mentioned that more research is needed to study these consequences. In addition, communication between online members may evoke negative emotions towards the company and an undesirable influence on consumers [2], some of which may also involve accidental problems. This study broadens the discourse on the value co-destruction and offers a more critical perspective on the experiences of Internet users in the virtual brand communities. More importantly, it aims to explore the causes, the emotions and the practices of the value co-destruction through netnography [10] and the method of critical incidents [4]. Our research offers a new perspective to practitioners for understanding consumer expectations and behavior. There is a need to re-evaluate and re-design value propositions in line with organizational capabilities and consumers' resource needs.

In the first part, we will focus on the literature on the value co-destruction. An empirical methodology follows in a second part. Finally, we will present the results of netnography and critical incidents as well as the theoretical and managerial implications of our study.

2 Theoretical Framework

Because of growing consumer empowerment and the proliferation of the technology of information and communication (TIC), the value co-creation has emerged [20]. Nevertheless, the value can be destroyed when TIC become integrated and used [13]. According to Neuhofer [13], the value co-creation may not always be positive, but also negative [15]. In addition, it may not even be created, but destroyed by the actors or resources that are integrated into the co-creation process of the value.

2.1 The Value

"*The definition of value is one of the most controversial problems in marketing literature*" [6, 7, 13]. According to Fernández and Bonillo (2006, p. 55), the value is "*a cognitive-affective evaluation of an exchange relationship carried out by a person at any stage of the process of purchase decision, characterized by a string of tangible and/or intangible elements which determine, and are also capable of, a comparative, personal, and preferential judgment conditioned by the time, place, and circumstances of the evaluation*". For their part, Vargo et al. [20] defined the value as an improvement in the welfare system [15, 18]. However, the value can not only be co-created, but it seems logically possible that it can be co-destroyed through the interactional processes. Specifically, the value can be co-destroyed through interactions between different service systems whether directly (person-to-person) or indirectly (via devices such as goods).

2.2 The Value Co-creation

The definitions of value co-creation, which are the subject of several researches, are not the same but share several common features and reveal particular characteristics from one definition to another. Thus, the first work focusing on the concept of value co-creation belongs to Vargo and Lusch [20] who have admitted that "the consumer is always a co-creator of value" [19]. This work of Vargo and Lush [20] is of great interest in leading to customer loyalty and unloading part of the costly, risky or low value-added creation cycle (cited by Dal Zotto and Vaujany 2011, p. 4).

Alongside firms that have increased their efforts to implement co-creation strategies (Dong et al. 2008, p. 124; Payne et al. 2008, p. 85, noted by Heidenreich et al. 2015, p. 281), consumer invests considerable effort and share feedback for a personalized experience (Etgar 2008, p. 98; Hoyer et al. 2010, p. 291). This means that there is an agreement that the firm and the consumer participate together in the creation of the common value (Etgar 2008, p. 98; Prahalad and Ramrtaswamy 2004, p. 7).

At the same time, it is important to note that consumer invests considerable time and effort in co-creating value. As described by Payne et al. (2009), the value co-creation implies that the value is determined once the product is consumed [19]. Then, the creator of the mark leaves a part of his power to the consumer so that the consumer can reinterpret the mark in his own way (Muniz and Schau 2005, p. 745) [19]. This is called "empowerment".

This new trend in the literature, which considers consumer not as a simple consumer of the firm's products, but as a co-creator of the value (Franke and Piller 2004, p. 402) [17] is increasingly evolving especially with the advent of virtual communities.

2.3 The Value Co-destruction

Created by the company during the production process, value can be destroyed by the consumer at the time of consumption [15]. From here appears the concept of the value co-destruction that emerged when the information is incomplete or misinterpreted [16] and whose activities and requirements are not aligned [8–12]. We cite in this connection the most popular definition, that of Plé and Chumpitaz Càceres [15], which defined the value co-destruction as «*an interactional process between service systems that results in a decline in at least one of the systems' well-being*». Also, the value co-destruction is although often described as the opposite of the value co-creation, considers practices that lead to a decrease in value for at least one of the actors (Lecrercq et al. 2016). The value co-destruction can occur intentionally or accidentally (involuntarily) through the misuse of resources leading to an overall decrease in the value [15]. This can be caused by an imbalance between the level of co-created the value of a company and that of its customer (Woodruff and Flint 2006). Smith (18) suggested that loss of resources is a key indicator of the value co-destruction and this contributes to negative emotional states which are used not only positively but also in a detrimental way. Let us cite the example of Plé and Chumpitaz Càceres [15], the consumer who buys a car but does not maintain it, destroys the value for himself. In addition, it destroys the value for the company that sold the product and harms its image by communicating its negative opinion to other people through negative word of mouth.

Previous studies have focused on co-destruction between service systems, however, in our study we focused on the value co-destruction in the virtual community. We will try, therefore, to study the causes, emotions and practices of the value co-destruction in the virtual community.

3 Methodology

We have adopted netnography as a qualitative research method as it is strongly associated with the emergence of virtual communities [10], and we have integrated the technique of critical incidents [18] in order to know the bad experiences of Internet users. To do this, we have selected three communities belonging to the telecommunication sector as research fields such as «TunisieTélécom», «OoredooTunisie» and «Orange». We followed Bernard's criteria for selection [3] when choosing these three communities, such as the proximity with our research question, the intensity of the traffic, the great number of these communities members (Appendix 1), the descriptive richness of data especially the negative comments and the large number of interactions between participants and the several negative comments. Six steps of the netnography recommended by Kozinets [2] were followed. At the first step, we entered in the social network "Facebook" and then in the three communities pages. Then we clicked on "I like" so that we could easily immerse our self in the group. The access to information was public (confidentiality: public) and the interactions are stored on the community and accessible to all visitors. In order to fully understand the culture of the three virtual communities, the rituals and the behaviors of the members, we asked for help from the creators of the communities since they had been immersed for a long time in the communities. Bernard [3] noted that it is necessary to ensure the reality of the virtual community and not just temporary gatherings. Thus, the three communities were co-created by the administrators of the pages and by its members. Secondly, to collect the data, we followed the five phases of Bernard [3] combined to the critical incidents technique. In a non-participating observation, we collected 92 negative comments. To illustrate, we have read the communications of members posted in the three communities by filtering the data. Bernard [3] postulated that we should not confuse between social communications (for example: "Hello, how are you?") and informational communications (for example: "I have problems with my Modem, it disconnects without stopping, how to change it ..."). This led us to choose well the textual data presented in the community. In addition, we must also differentiate communications in relation to the community's focus, that is, the activity of consumption that is the subject of study and those that are not relevant [3]. To this end, we have limited ourselves to informational communications and which are related to our study. Since we are in the first phase of data collection as the non-participant observation, we have copied and translated the comments without interacting with the members, in addition we have preferred to keep the confidentiality of the Identity of members. As part of a participatory observation, groups of discussions and online interviews were conducted using a guide of interview (Appendix 2). Referring to Bernard's recommendations [3], we created an online discussion group on the social network "Facebook", and then we invited the members already immersed in their community to discuss with them. We asked these members about their gender, age and occupation. Since we received few responses in the focus group, we sent individual messages to each member by using our interview guide. A critical incident technique was adopted where Internet users reported their negative experiences. Given the lack of documentation on the properties of the concept and the need to identify and classify them, the critical incident technique (CIT) would be an appropriate method of research [4] to collect descriptions of experiments [23]. This is due to the fact that, in the case of the virtual community, external events, such as critical incidents, trigger new information. When collecting data, we followed the two sampling criteria or principles of diversification and saturation. Then, we have applied a content analysis. Referring to the recommendations of Decrop and Degroote (2015, p. 7), we have systematically coded our data by an open coding (an inductive coding applied to the first reading). Given that the themes identified by open coding deserve closer examination (Hollebeek and Chen 2014, p. 65) and a link to our research question, we have switched to axial coding (a grouping of codes in large themes) before confronting them in a logic of comparison and synthesis (selective coding). Indeed, the coding of the data serves to identify variables that reflect the behavior of the participants. To do this, we used the "Lexico 3" software which we were able to know the repeated segments and the frequency of repeated words. Subsequently, we were able to classify themes into categories in a theme grid and extract from the verbatim of each member the necessary themes. Our goal is to identify the emerging themes of the content of three virtual communities.

4 Causes, Emotions and Practices of the Value Co-destruction: Key Findings

Content analysis allowed us to identify the causes, emotions and practices of the value co-destruction in the virtual brand community.

4.1 The Causes of the Value Co-destruction

Internet users exchange information about their purchases and consumer experiences with other consumers (Weiss et al. 2008) [2]. Nevertheless, these experiences may be negative. From this point of view, the method of critical incidents has revealed that members have reported the failure of their experiences with telecommunications brands and in particular the lack of response by the company to their complaints. «*Really exceptional service every month the same problem to buy a package every day I call the customer service and the right answer one will pass the complaint to the technical service», «Still no answer????».* These results are consistent with the study of Prior and Marcos-Cuevas [16] who postulated that the value co-destruction is the result of interactions and evaluations of non-positive experiences among actors. Moreover, the absence of the firm's credibility can be a primary cause of the value co-destruction, for example: «it's cheating and orange is not honest with its customers». These same Internet users added that the misbehavior of the company is a primary cause of the value co-destruction in the virtual community. We can relate all these bad practices to the corporate misbehavior including its injustice, non credibility, scam, theft, cheating, shame and opportunistic behavior. In the same context and in accordance with the results of Wellman (1997, p. 191), we observed that many comments were published by anonymous, which also reduces the credibility of the information. What we found from netnography is consistent with the study by Echeverri and Skalen (2011) who found that the value co-destruction occurs when information is incomplete or misinterpreted or when activities and requirements of the actors are not aligned "I received on 26/10 a call from one of your agents to inform me that the problem was solved when it was not the case".

4.2 The Emotions of the Value Co-destruction

According to Adjei et al. [2], communicating with other online members evokes positive emotions in a community member. While we agree with their assertion, we argue that online communication between members may also evoke negative emotions toward the company. Our qualitative study allowed the emergence of different negative emotions that could destroy the value in the virtual community. Following the proposal of Adjei et al. [2], in their future paths of research, which examines whether negative emotions can be evoked as a result of social media communications, we find out from the netnography that members are very dissatisfied and very angry, we quote: "I entered Ooredoo to express my anger towards the brand." So, they do not just declare their negative affections to the company, but also their dissatisfaction with the brand. At the same time, they acknowledged their discontent and hatred for the brand's services, for example: "I do not like Ooredoo because it is raising price values and quality is mediocre, "Worst connection in the world". The researchers (i.e. Bechwati and Morrin 2003; Gregoire and Fisher 2008) suggested that these feelings are motivated by lack of equity in the company. As a result, the majority of members stated their negative impressions of telecommunications brands, as demonstrated by Adjei et al. [2], the public nature of social networks accentuates, on the one hand, the emotional reaction of disappointment: "I am really disappointed", and on the other hand, the feeling of betrayal of Internet users. Indeed, by discovering on social networking sites that another consumer receives preferential treatment will evoke feelings of betrayal as well as the desire for retaliation or revenge that publicly received information through social networking sites raise levels of betrayal higher than information received via non-social networking platforms. Grégoire and Fisher (2008) defined betrayal as "a motivation that forces clients to restore fairness through retaliatory behavior such as complaint or commutation". In fact, this negative emotion has arisen when the consumer becomes aware that other consumers are receiving benefits, which he himself is not. Members can even take pleasure in the misfortune of others (Hickman and Ward 2007) to the extent that a consumer can inform others about the free upgrades he has received from the company.

4.3   The Practices of the Value Co-destruction

Many practices can destroy the value in the virtual community. Adjei et al. [2] reported that these practices are unintentional or accidental, or even unplanned negative practices that occur when users communicate with one another via social networking sites. First, older members may discourage and delay participation in the new subscriber community as they have articulated poor expectations through the spread of negative word-of-mouth. They have even reduced the attitude of new members "This is your first month of package you will see for the rest personally I have terminated and 80% of my contacts have changed supplier." Moreover, consumers have made unfair comparisons with other competing brands, we quote some comments illustrating this point of view: "I sincerely think about changing the supplier to Orange or Tunisie Telecom", "if not I try to change the operator". Muñiz and Hamer(2001) and Muñiz and O'Guinn (2001) have shown that they are loyal to a competing brand. This can lead to a decline in product sales and resistance to brand use. Adjei et al. [2] defined the reduction of patronage as reducing the frequency of purchases and visits and increasing the frequency of purchases from competitors (De Wulf et al. 2001; Grégoire and Fisher 2006). Similarly, members can express their loyalty to the brand by their opposition to competing brands (Thompson and Sinha 2008, [2]). More importantly, members of the community can not only launch their events to promote their brand. At the same time, instead of advising other members on how to approach the market, members strongly advised other members to buy the brand. On the other hand, the members behaved negatively, for example, by laughing at the "Ooredoo" brand: "Hi it goes by costs 2000 if I call abroad, it will cost less." In addition, qualitative analysis has informed us of the most dangerous behaviors leading to the value co-destruction as well as the complaint «they are thieves, you must complain». «I will throw the chip and buy another from another supplier».

5   Conclusion

5.1   Theoretical and Managerial Implications

Our study resulted in certain numbers of theoretical and managerial implications. Using the qualitative method of netnography with the critical incident method, we clarified for both managers and academics that the value co-creation is not the only possible outcome of interaction between actors, but also, the value can be co-destroyed by the interaction between the supplier and the consumers. Thus, the possibility of co-destruction of the interactional value should not be neglected. We first explored the causes of the value co-destruction, notably the failure of consumer experiences with the brand because of the bad behavior of the company, and secondly, the emotions. Next, we proposed that several practices could destroy the value in the virtual brand community. Thus, by exploring the causes, emotions and practices of the value co-destruction, this study extends current knowledge into the marketing literature. In an effort to reap the benefits of social networking sites, marketing managers can unwittingly create unwanted outcomes for the company. As such, we advise marketers of social networking sites dedicated to the purchase and consumption of products to become aware of this undesirable consequence.

5.2   Limitations and Future Research

Despite all efforts to complete this study, a number of limitations need to be highlighted. To these limits, we will associate future paths of research. In the first place, we have limited ourselves to a qualitative exploratory study using netnography and a technique of critical incidents, and a quantitative phase will be necessary. A second limitation is in choosing a single online environment such as telecommunication which our study may not be representative of all the communities developed on the Facebook social network. Thus, future research should investigate other categories of products and communities in order to have a rich understanding about the value co-destruction. A third limitation concerns the longitude of the study whose practices of the value co-creation are dynamic and their effects will change with time. Therefore, a longitudinal and in-depth study will be necessary in order to follow the changes in the uses that the members make.

Appendix 1: The Number of Communities Members at 2017

The community	The number of community members
«Tunisie Télécom»	995.145
«Ooredoo Tunisie»	2.306.360
«Orange»	22.595.153

Appendix 2: The Guide of Interview

Theme 1: The causes of value co-destruction
1/ Why do you behave negatively in the community?
2/ How do you evaluated your experience with the brand and the community?
Theme 2: The emotions of value co-destruction
3/ What are your emotions towards the community and the brand?
4/ Do you think you're satisfied with the community which you participated in the creation?
Theme 3: The practices of value co-destruction
5/ How do you behave in the community?
6/ What are the consequences of your comments on the company?

Appendix 3: Characteristics of the Sample

	Members profile	Number of members publishing negative comments in percent (%)
Gender	Man	66,3%
	Woman	33,7%
Age	Between 15–25	58,3%
	Between 26–30	16,7%
	Between 30–40	25%
Socio-professional Category	Pupil	10,3%
	Student	31%
	Unemployed	6,9%
	Liberal profession	6,9%
	Employee	44,9%

Appendix 4: The Causes of Value Co-destruction

Les causes de la co-destruction de la valeur	«Verbatims»
Corporate Misbehavior: – The lack of credibility of the information and the company – The scam, the flight, the cheating and the shame of the company – Opportunistic behavior – The lack of the response by the company to consumer complaints. – The unfairness – The lack of the transparency of the company	• *«c'est de la triche et orange n'est pas honnête avec ses clients»* • *«Tunisie Télécom tout simplement essaie de m'arnaquer»*, *«tout le monde souffre de l'arnaque de Ooredoo»*, *«oui ooreedoo Tunisie est devenue une firme d'arnaque»* • *«c'est la triche vous volez»*, *«cessez de voler, c'est la honte»*, *« vous n'aimez que le vol»*, *«depuis que vous avez changé votre nom vous avez perdu votre crédibilité»* • *«pourquoi vous mentez»* • *«Toujours pas de réponse????»* • It's cheating and orange is not honest with its customers" • "Tunisie Telecom simply tries to rip me off", "everyone suffers from the rip-off of Ooredoo", "yes ooreedooTunisie has become a rip-off firm" • "It's the cheat, you steal", "Stop stealing, it's a shame", "You only like stealing", "Since you changed your name you have lost your credibility" • "whyyou lie" • "Still no answer????"
– The failure of the consumer experiences with the brand	• *«j'ai rechargé 5 dt ils m'ont volé 2 dt forfait»* • *«Quand j'achète un forfait Facebook je retourne après un jour je le trouve terminé. C'est illicite»* • *«chaque mois le même problème pour acheter un forfait chaque jour j'appelle le service client et la bonne réponse on va passer la réclamation au service technique»* • *«Ya deux mois quand je recharge une carte jamais je la trouve complète par exemple quand je recharge 1 dt je reçois que 500 mil»* • *«j'ai dépassé les 9dt de consommation mais je ne trouve pas 1Go d internet?!»* • *«Honnêtement c'est pas la première fois que j'ai essayé de recharger mon numéro et ça passe dans le vide exp aujourd'hui j'ai rechargé 2 dt et soudain ça passe dans le vide c'est dommage comme même à noter bien que j'ai bénéficié de l'offre de kriddi qu'une seule fois. Même chose pour l'offre flexi j'ai 600 mo et se passe dans le vide j'ai connecté que pour une demi-heure»* • *«j'ai acheté un forfait 2dt/moisil a resté 5 jrs puis il a terminé???»* • "I recharged 5 dt, they stole 2 dt forfeit" • "When I buy a Facebook package I return after one day I find it finished. It's illegal" • "every month the same problem: to buy a package every day I call the customer service and the right answer: we will pass the complaint to the technical department" • "Two months ago when I refilled a card. Never I find it complete for example when I recharge 1 dt I get 500 m" • "I have exceeded the 9 dt of consumption, but I don't find 1Go of Internet!" • "Honestly this is not the first time I tried to refill my prepaid card and but my card was not credited. For example, today I recharged 2 dt, but my card was not credited, that's too bad. To note that I have benefited from the offer of

(continued)

(continued)

Les causes de la co-destruction de la valeur	«Verbatims»
	kriddi only once. Same thing for the flexi offer: I have 600 Mo and happens in the vacuum. I only logged in for half an hour" • "I bought a 2dt/month package. It stayed 5 days then it finished??? "
– Conflicts between members – Negative interactions between the company and the members – The imbalance between the value co-created by the company and the value co-created by the user – Loss of resources	• «Je vous admire Que Dieu vous aide», M.M. a répondu: «Arrête!!!! Ooredoo est voleuse» • I admire you God helps you, "Mr. M. answered: " Stop!!!! Ooredoois a thief"
– Incomplete or misinterpreted informations	• «bonsoir, j'ai fait un contrat avec orange concernant la clé premium illimité de 08 h à 19 h et 15 Go de 19 h à 08 h et maintenant ils me disent que l'illimité est " limité " à 25 Go et ma connexion internet a cessé le 26 octobre 2016 et quand j'ai contacté le service clientèle ilsm'ont dit "nous s'excusons"» • «j'ai reçu le 26/10 un appel d'un de vos agents pour m'informer que le problème a été résolu alors que ce n'était pas le cas» • "Good evening, I made a contract with orange regarding the unlimited premium key from 08 h to 19 h and 15Go from 19 h to 08 h. Now they tell me that unlimited is "limited" to 25 GB and my internet connection has stopped October 26, 2016 and when I contacted the customer service they said "we apologize" • "I received a call from one of your agents on 26/10 to let me know that the problem was solved when it was not"
– The change in the name and nationality of the brand	• «pourquoi vous avez changé le forfait fb à l'ancien était plus mieux» • «car quand elle était sous le nom du Tunisiana elle était le leader parmi ses concurrents et on perçoit les offres sur la page communautaire mais quand elle a changé son nom à ooreedoo, plusieurs ont migré vers Orange» • «à l'ancien, vous êtes la meilleure firme et maintenant vous êtes trop chers, nous détestons ce nouveau service» • «nous voulons reprendre l'ancienne Tunisiana» • "Why you changed the fbpackage. The former was better" • "because when she was under the name of Tunisiana she was the leader among her competitors and we see the offers on the community page but when she changed her name to ooreedoo, several migrated to Orange" • "to the old, you are the best firm and now you are too expensive, we hate this new service" • "we want to take over the old Tunisiana"

Appendix 5: The Emotions of Value Co-destruction

Les émotions de la co-destruction de la valeur	«Verbatims»
– The anger	• «Je suis entré à Ooredoo pour exprimer ma colère envers la marque» • "I entered Ooredoo to express my anger towards the brand"
– The unfavorable impression on the brand – The negative affect on the company – The negative evaluation – The dissatisfaction	• «j'aime pas Ooredoo puisqu'elle est en train d'augmenter les valeurs de prix et la qualité est médiocre» • «on n'aime pas», «j'ai détesté: votre forfait est épuisé», «on a détesté ooredoo», «depuis quand Ooredoo fait du bien!» • I dont like Ooredoo since it is increasing the prices and the quality is poor • "we don't like", "I hated: your package is exhausted", "We have detested Ooredoo", "Since when Ooredoo does good!"
– The feeling of treason	• R.M: «Vous mentez! je joue presque chaque jour et jamais que je gagne». N.S a répondu: «Ce n'est pas vrai moi j'ai gagné un smart phone la semaine précédente» • R.M: "You are lying! I play almost every day and never win" N.S replied, "It's not true, I won a smart phone last week"
– The disappointment – The frustration	• «je suis vraiment déçu» • I'mreallydisappointed
– The hostility	• «Ma relation avec la marque est très mauvaise» • «la plus mauvaise connexion dans le monde» • «La situation est devenue critique et gênante» • "My relationship with the brand is very bad» • "the worst connection in the world" • "The situation has become critical and embarrassing"

Appendix 6: The Practices of Value Co-destruction

Les pratiques de la co-destruction de la valeur	«Verbatims»
– The negative word of mouth – The discouragement of community participation – The articulation of bad behavioral expectations – The publication of negative messages about the brand – The bad governance of the consumer on the company	• «Vous êtes incapables» • «c'est ton premier mois de forfait tu verras pour le reste personnellement j'ai résilié et 80% de mes contacts ont changé le fournisseur» • "You are incapable" • "This is your first month of package you will see for the rest, personally I have canceled and 80% of my contacts have changed the supplier"
– Not recommending the other members to buy the brand – Reduce the attitude of new members of the community	•«c'est ton premier mois de forfait tu verras pour le reste personnellement j'ai résilié et plein de mes contacts aussi 80% ils ont changé le fournisseur» • «je vais jeter la puce et acheter une autre depuis un autre fournisseur»

(continued)

(continued)

Les pratiques de la co-destruction de la valeur	«Verbatims»
	• «si non j'essaye de changer l'opérateur» • «je veux résilier ma ligne» • "This is your first month of package you will see for the rest personally I canceled and full of my contacts also 80% they changed the supplier" • "I will throw the chip and buy another from another supplier" • "if not I try to change the operator" • "I want to cancel my line"
– The boycott of the brand – The collective complaints	• «Le boycott du grand voleur Ooredoo» • "The boycott of the great thief Ooredoo"
– The complaint	• «ce sont des voleurs, tu dois plaindre» • "They are thieves, you must pity"
– The reduce of patronage – The switching – Reducing the use of the brand – The reduce of the sales of products	• «je réfléchis sincèrement à changer le fournisseur vers Orange ou Tunisie Télécom» • «prochainement avec la portabilité des numéros on peut changer l'opérateur Ooredoo par TT ou Orange en gardant le même numéro» • «j'ai déjà acheté une puce orange moi et toute la famille» • «quand j'avais un mobile wifi télécom jamais passé avec moi ça» • «je vais jeter la puce et acheter une autre depuis un autre fournisseur» • "I sincerely think about changing the provider to Orange or Tunisie Telecom" • "soon with the portability of the numbers one can change the operator Ooredoo by TT or Orange keeping the same number" • "I already bought an orange chip me and the whole family" • "When I had a mobile wifi telecom never spent with me" • "I will throw away the chip and buy another from another supplier"
– The revenge – The threat	• «je vais exploser la plus proche agence de vous si vous ne me récupérez pas mes argents» • "I will explode the nearest agency of you if you do not recover my money"
– The mock of the brand	• «Slt cv bye coûte 2000 😊 si j'appelle l'étranger, elle va coûter moins cher» • "Slt cv bye costs 2000 😊 If I call abroad, it will cost less"
– Narration of bad stories about the brand	• «j'ai rechargé 5 dinars ils m'ont volé 2 dinars forfait et moi je ne l'aime plus!!! Quelle vole???? arrêtez de voler» • "I reloaded 5 dinars they stole 2 dinars forfait and I do not like it anymore!!! What theft???? Stop stealing"

References

1. Algesheimer, R., Dholakia, U.M., Herrmann, A.: The social influence of community: evidence from European car clubs. J. Market. **69**, 19–34 (2005). doi:10.1509/jmkg.69.3.19.66363

2. Adjei, M.T., Nowlin, E.L., Ang, T.: The collateral damage of C2C communications on social networking sites: the moderating role of firm responsiveness and perceived fairness. J. Market. Theory Pract. **24**(2), 166–185 (2016)
3. Bernard, Y.: La netnographie: une nouvelle méthode d'enquête qualitative basée sur les communautés virtuelles de consommation. Décis. Market. **36**, 49–62 (2004)
4. Bitner, M.J., Booms, B.B., Mohr, L.A.: Critical service encounters: the employee's viewpoint. J. Market. **58**(1), 95–106 (1994)
5. Carlson, B.D., Suter, T.A., Brown, T.J.: Social versus psychological brand community: the role of psychological sense of brand community. J. Bus. Res. **61**, 284–291 (2008)
6. Day, E.: The role of value in consumer satisfaction. J. Consum. Satis. Dissatis. Complaining Behav. **15**, 22–32 (2002)
7. Day, E., Crask, M.R.: Value assessment: the antecedent of customer satisfaction. J. Consum. Satis. Dissatis. Complaining Behav. **13**, 52–60 (2000)
8. Echeverri, P., Skalen, P.: Co-creation and co-destruction: a practice-theory based study of interactive value formation. Market. Theory **11**(3), 351–373 (2011)
9. Gebauer, J., Füller, J., Pezzei, R.: The dark and the bright side of co creation: triggers of member behavior in online innovation communities. J. Bus. Res. **66**, 1516–1527 (2013)
10. Kozinets, R.V.: Marketing netnography: a new research method. Methodol. Innov. Online **7**(1), 37–45 (2012). York University, Toronto, Canada
11. Laroche, M., Habibi, M., Richard, M., Sankaranarayanan, R.: The effects of social media based brand communities on brand community markers, value creation practices, brand trust and brand loyalty. Comput. Hum. Behav. **28**(5), 1755–1767 (2012)
12. Luo, N., Zhang, M., Liu, W.: The effects of value co-creation practices on building harmonious brand community and achieving brand loyalty on social media in China. Comput. Hum. Behav. **48**, 492–499 (2015)
13. Majboub, W.: Co-creation of value or co-creation of experience? Interrogations in the field of cultural tourism. Int. J. Saf. Secur. Tour. **7**, 12–31 (2014)
14. Neuhofer, B.: Value co-creation and co-destruction in connected tourist experiences (2016)
15. Plé, L., Chumpitaz Cáceres, R.: Not always co-creation: introducing interactional codestruction of value in service-dominant logic. J. Serv. Market. **24**(6), 430–437 (2010)
16. Prior, D., Marcos-Cuevas, J.: Value co-destruction in interfirm relationships: the impact of actor engagement styles. Market. Theory 1–20 (2016)
17. Schau, H.J., Muñiz, J.A., Arnould, E.J.: How brand community practices create value. J. Market. **73**, 30–51 (2009)
18. Smith, A.M.: The value co-destruction process: a customer resource perspective. Eur. J. Market. **47**, 1889–1909 (2013)
19. Susilo, C.: Communauté de marque et revitalisation de marque: le cas de Polaroid. Mémoire de recherche. HEC Montréal, Canada, pp. 1–130 (2012)
20. Vargo, S.L., Lusch, R.F.: Evolving to a new dominant logic for marketing. J. Market. **68**, 1–17 (2004)
21. Vargo, S.L., Maglio, P.P., Akaka, M.A.: On value and value co-creation: a service systems and service logic perspective. Eur. Manage. J. **26**(3), 145–152 (2008)
22. Vafeas, M., Hughes, T., Hilton, T.: Antecedents to value diminution: a dyadic perspective. Market. Theory **16**(4), 469–491 (2016) ISSN 1470–5931
23. Van Doorn, J., Verhoef, P.C.: Critical Incidents and the Impact of Satisfaction on customer Share. J. Market. **72**(4), 123–142 (2008)
24. Zaglia, M.E.: Brand communities embedded in social networks. J. Bus. Res. **66**, 216–223 (2013)

Empirical Study of Algerian Web Users' Behavior

The Case of Ouedkniss.Com

Fares Medjani[✉]

HEC Alger, Tipaza, Algeria
famedjani@gmail.com

Abstract. ICT's are changing business and marketing, and internet is a major evolution that leads to electronic commerce. In Algeria, e-commerce is perceived as "Myth" for some while it is a "Reality" for others. Being visited more than Facebook in Algeria, Ouedkniss.com this platform of classified ads is an exception of a website that succeeded in the Algerian market. Thus, understanding the determinants of its adoption is essential for Algerian managers to launch successful e-commerce website.

The current study was supported by a quantitative research using an online survey to explore the role of the variables of the Technology Acceptance Model in the adoption of an e-commerce website. After validating the measurement scales used in the study, the results confirmed that TAM variables are relevant to explain the adoption of Ouedkniss.com and e-commerce. Moreover, the determinants of perceived usefulness were significant except one variable. Finally, the determinants of perceived ease of use were not significant.

Thus, this model can be used, taking in consideration the significance of all variables, to explain and predict the adoption of e-commerce by Algerian users.

Keywords: E-commerce · Electronic commerce · Online consumer behavior · Technology Acceptance Model

1 Introduction

The development of new Information and Communication Technologies (ICT's) has affected the human behavior and deeply changed the business environment. It is considered as a Deadly Marketing Sin the firm that has not made maximum use of technology (Kotler 2004). The emergence of e-commerce is considered as the truly revolutionary impact of the internet revolution (Drucker 2002, pp. 3–4).

In Algeria, a country dominated by informal economy, e-commerce is perceived as "Myth" for some while it is a "Reality" for others (Makhlouf and Belattaf 2013). The first steps begin to appear in the market: few e-commerce websites, online reservation, billing and more recently the launching of electronic payment. However, many barriers still block the business use of ICT's more particularly the level of investments in technology and the lack of trained managers (Lamiri 2016). Experts agree that there is a gap to gain in Digital Economy (Grar 2016; Kahlane 2016; Lamiri 2016).

© Springer International Publishing AG 2017
R. Jallouli et al. (Eds.): ICDEc 2017, LNBIP 290, pp. 55–63, 2017.
DOI: 10.1007/978-3-319-62737-3_5

An exception occurs with the website Ouedkniss.com, this platform of classified ads is visited more than Facebook in Algeria (Alexa 2016). This success story did not lead to scientific research despite the added value of understanding its visitor's behavior. While, to the best of our knowledge, scientific research studied neither Algerian web users' behavior nor Ouedkniss case study, the purpose of this paper is to give a first exploratory study of the determinant of the adoption and use of this website based on the Technology Acceptance Model.

2 Context

Understanding the factors explaining the success of Ouedkniss.com is a starting point to succeed in the launch of e-commerce websites. This is the challenge for the Algerian managers: elaborating adapted e-commerce strategies for the Algerian market to meet potential customers' expectations. From this challenge we formulated the following research problem: **what are the variables that determine the adoption of e-commerce by Algerian web users?**

In fact, since the last years of the 1980's, acceptation and use of technologies was an interesting topic in scientific research (Ben Boubaker 2013, p. 1) and many scientific articles and theories appeared (Yousafzai et al. 2010) as Theory of Planned Behavior – TPB (Ajzen 1985, 1991, 2011); Innovations Diffusion Theory - IDT (Roger 2003); and recently the Unified Theory of Acceptance and Use of Technology – UTAUT (Venkatesh et al. 2003).

From this view, the Technology Acceptance Model – TAM can provide us an answer, and explain the adoption and use of an Information System or any other information support. This model shows that the perception (of usefulness and ease of use) influences positively the attitude toward the use. This last, influences directly the use of an information system (Ouedkniss and e-commerce in our case study).

2.1 Origin of TAM and Evolutions

Based on the work of "Fishbein et Ajzen, Theory of Reasoned Action" the first TAM model was developed in a PhD dissertation (Davis 1986).

The first modified version of the TAM appeared by adding a new variable to the model "Behavioral Intention to Use" (Davis et al. 1989) and this model was cited 17.102 times (Scholar Google 2016). This confirms the importance of this model in scientific literature.

A new study conducted on 107 users lead to a new modification in the model, following the correlations and the significant impacts the "Attitude" was eliminated (Venkatesh and Davis 1996) (Fig. 1).

Due to its simplicity the TAM was the most attractive model compared to Theory of Reasoned Action and Theory of Planned Behavior (Chuttur 2009).

Measuring attitude should be done through its three components or by specifying which of the three is of focal concern (Breckler 1984; Rosenberg and Hovland 1960). Thus, to measure "Behavioral Intention" which refers to attitude we will use the "conative component of attitude".

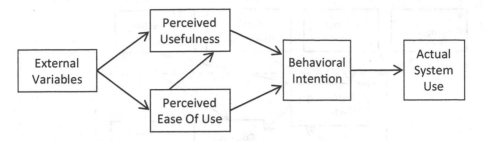

Fig. 1. Final version of TAM (Venkatesh and Davis 1996, p. 453).

Venkatesh and Davis (2000) extended the TAM to TAM2 by identifying the variables influencing the perceived usefulness. They proposed the following variables: subjective norm, image, job relevance, output quality and result demonstrability. Note that as Ouedkniss.com is not used for work, "job relevance" is excluded from our research framework. The variables influencing the perceived ease of use were highlighted by Venkatesh (2000), his study concluded that: computer self-efficacy, perception of external control, computer anxiety, computer playfulness, perceived enjoyment and objective usability are influencing directly the perceived ease of use. "Objective usability" is measured by comparing time spent by an expert and time spent by the respondents. As this variable can't be measured using a survey, it is excluded from our research framework. Combining the antecedents of perceived usefulness and perceived ease of use was the last evolution to TAM3 Model (Venkatesh and Bala 2008). This lead us to the following hypotheses:

$$\text{H1:PU} = \beta_{01} + \beta_{11}\text{SN} + \beta_{21}\text{IMG} + \beta_{31}\text{OUT} + \beta_{41}\text{RD} + \beta_{51}\text{PEU} + \varepsilon_1 \qquad (1)$$

$$\text{H2:PEU} = \beta_{02} + \beta_{12}\text{CE} + \beta_{22}\text{PEC} + \beta_{32}\text{CA} + \beta_{42}\text{CP} + \beta_{52}\text{PE} + \varepsilon_2 \qquad (2)$$

$$\text{H3:ATT} = \beta_{03} + \beta_{13}\text{PU} + \beta_{23}\text{PEU} + \beta_{33}\text{SN} + \varepsilon_3 \qquad (3)$$

$$\text{H4:U} = \beta_{05} + \beta_{15}\text{ATT} + \varepsilon_5 \qquad (4)$$

* Abbreviations	OUT: Output Quality
PU: Perceived Usefulness	RD: Results Demonstrability
PEU: Perceived Ease of Use	CE: Computer self-efficacy
ATT : Attitude (conative)	PEC: Perception of External Control
U: Use	CA: Computer Anxiety
SN: Subjective Norm	CP: Computer Playfulness
IMG: Image	PE: Perceived Enjoyment

These hypotheses are illustrated in the research framework below (Fig. 2).

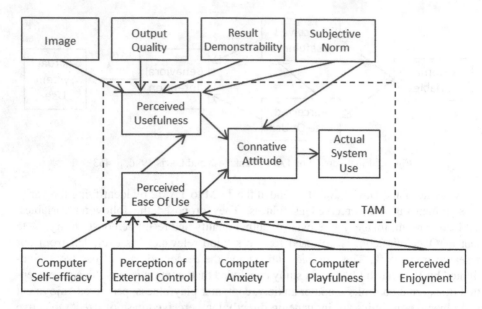

Fig. 2. Research Framework adapted from Venkatesh and Bala (2008)

3   Methodology

To answer our research problem and test our hypothesis, we opted for the analytical research methodology with a quantitative research using survey and statistical techniques. The Table 1 below summarizes the survey information.

Table 1. Summary of the survey

Target population	Ouedkniss.com users. As it is difficult to manipulate the sample online the questionnaire contains a filter question to filter users from non-users
Sampling method	As sampling frame is not available, we used Convenience Sampling
Sample size	322 completed questionnaires 300 accepted and 22 rejected
Survey driving	Administration mode: Online survey It is the adequate mode for this study as the target population is online Length: 20 days
Questionnaire test	Before launching the Survey, the questionnaire was tested on 15 individuals. No one had difficulties to understand and answer the questions
Questionnaire items	The measurement scales used in this survey are inspired by the different works on TAM Model (Davis 1989, 1993; Venkatesh and Davis 2000; Venkatesh 2000)

4   Analysis and Results

Before testing our model, we start by an exploratory and confirmatory factor analysis (Gorsuch 1998; Thompson 2004) to test the validity & reliability of the measurement scales used in the survey (Hair et al. 2006).

Validity is tested with convergent validity; we use the Correlation Coefficient to test the correlation between the items used to measure the same variable. We also use Average Variance Extracted (AVE) (Malhotra and Dash 2011, p. 702). The Reliability of the scales is estimated by using the Cronbach's Alfa coefficient following the Table 2 values and the results are shown in Table 3.

Table 2. Cronbach's Alfa coefficient values (Carricano et al. 2010, p. 53).

<0.6	0.6 to 0.65	0.65 to 0.7	0.7 to 0.8	0.8 to 0.9	>0.9
Insufficient	Weak	Acceptable	Good	Very good	Reduce item

Table 3. Summary of validity and reliability tests

Variables	R	AVE	Cronbach's Alpha
Perceived utility	item1–item2: 0.630	0.760	0.842
	item1–item3: 0.660		
	item2–item3: 0.633		
Perceived ease of use	item1–item2: 0.190	3 items: 0.615	3 items: 0.699
	item1–item3: 0.755	2 items: 0.877	2 items: 0.860
	item2–item3: 0.337		
Image	item1–item2: 0.607	0.799	0.754
Output quality	item1–item2: 0.563	0.780	0.720
Result demonstrability	item1–item2: 0.616	0.808	0.761
Computer anxicty	item1–item2: 0.529	0.741	0.691

Following these results, the item2 of perceived Ease of use will not be considered in the analysis. Note that the variables which are not stated in the Table 3 were measured with a single item.

To test the coherence of our research model and capture the relationships between its variables, we use the structural equation modeling (SEM) with a PLS analysis which is the appropriate technique for validating models (Lowry and Gaskin 2014). The Table 4 summarizes the results, the relationships among variables and the hypothesis test.

These results are illustrated in the following (Fig. 3).

5 Discussion

Several findings emerge from our study. First, the results support a significant support for TAM model. The subjective norm, perceived usefulness and perceived ease of use explain more than 40% variance in conative attitude. Moreover, the conative component of attitude has a significant effect on the use and these conclusions are in agreement with the literature.

The results support the significance of the determinants of perceived usefulness too and are in accordance with previous research as 60% of the variance of perceived usefulness is explained. However, the effect of image is not significant even if the measurement items showed a good reliability.

Table 4. Summary of path coefficients and significance levels

Tested hypothesis relationships	R^2	βni	t student	Hypothesis test
H1-1. Subjective Norm → Perceived Usefulness	0.625	0.141	2.860**	Confirmed
H1-2. Image → Perceived Usefulness		−0.050	0.992	Rejected
H1-3. Output Quality → Perceived Usefulness		0.347	6.307***	Confirmed
H1-4. Demonstrability of Results → Perceived Usefulness		0.208	3.200**	Confirmed
H1-5. Perceived Ease of Use → Perceived Usefulness		0.290	4.687***	Confirmed
H2-1. Computer Self-efficacy → Perceived Ease of Use	0.396	0.161	1.640	Rejected
H2-2. Perception of External Control → Perceived Ease of Use		0.178	1.754	Rejected
H2-3. Computer Anxiety → (-) Perceived Ease of Use		−0.032	0.660	Rejected
H2-4. Computer Playfulness → Perceived Ease of Use		0.095	1.271	Rejected
H2-5. Perceived Enjoyment → Perceived Ease of Use		0.355	5.455***	Confirmed
H3-1. Perceived Usefulness → Attitude	0.413	0.305	5.297***	Confirmed
H3-2. Perceived Ease of Use → Attitude		0.303	4.789***	Confirmed
H3-4. Subjective Norm → Attitude		0.154	3.000**	Confirmed
H4. Attitude (conative) → Use	0.033	0.181	3.170**	Confirmed

*p < 0.05
**p < 0.01
***p < 0.001

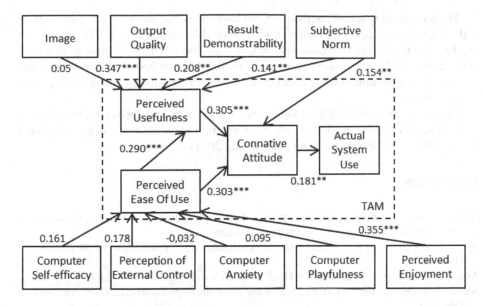

Fig. 3. The relationships between the research model variables

Finally, in contradiction with the literature, even if they explain around 40% of variance in the perceived ease of use, the antecedents of perceived ease of use were not significant except perceived enjoyment.

We consider this study as a first step in understanding Algerian web user's behavior by modeling and empirically testing the TAM in addition to the determinants of perceived ease of use and perceived utility.

One major theoretical contribution of our study is testing and validating empirically the TAM model in the context of Algerian web user's behavior as this model was generally tested in an Anglo-Saxon environment. From this view, the significance of the results shows that this model can be applied.

Furthermore, the present study findings have important managerial implications. In fact, it provides Algerian managers with variables to take in consideration when developing websites; the focus should be on designing useful websites which are easy to use.

6   Conclusion

To explain the behavior of using e-commerce, based on the Technology Acceptance Model, we applied our conceptual model to Ouedkniss.com users; the results confirmed that TAM variables are relevant to explain the adoption of this website. In addition, investigating the determinants of perceived usefulness showed their significant effect except the image. Finally, in contradiction with literature the variables explaining perceived ease of use were not significant except the perceived enjoyment. Thus, this model can be used, taking in consideration the significance of all variable, to explain and predict the adoption of e-commerce websites by the Algerian web users.

The present study was not without several limitations that should be cited. In addition to the limitations of the online administration mode (Blasius and Brandt 2010; Heerwegh and Loosveldt 2008) and the convenience sampling, the TAM has limitations in testing methodology and in relationships within the model (Chuttur 2009). Furthermore, some constructs were measured using a single item, thus, theirs validity and reliability can't be estimated.

Future research can study the adoption of the e-payment launched recently in Algeria by both sellers and buyer, or to study other external variables such as organizational support and social pressure (Chang and Cheung 2001).

References

Ajzen, I.: From intentions to actions: a theory of planned behavior. In: Kuhl, J., Beckmann, J. (eds.) Action Control, pp. 11–39. Springer, Heidelberg (1985)

Ajzen, I.: The theory of planned behavior. Organ. Behav. Hum. Decis. Process. **50**(2), 179–211 (1991)

Ajzen, I.: Theory of planned behavior. Handb. Theor. Soc. Psychol. **1**, 438 (2011)

Alexa: Top Sites in Algeria (2016). http://www.alexa.com/topsites/countries/DZ. Accessed 15 Dec 2016

Ben Boubaker, K.: Technology characteristics and IT acceptance and use: development and validation of a research model. Ph.D. thesis. HEC Montréal (2013)

Blasius, J., Brandt, M.: Representativeness in online surveys through stratified samples. Bulletin de Méthodologie Sociologique **107**(1), 5–21 (2010)

Breckler, S.: Empirical validation of affect, behavior, and cognition as distinct components of attitude. J. Pers. Soc. Psychol. **47**(6), 1191–1205 (1984)

Carricano, M., Poujol, F., Bertrandias, L.: Analyse de données avec spss®. Pearson Education, France (2010)

Chang, M., Cheung, W.: Determinants of the intention to use Internet/WWW at work: a confirmatory study. Inf. Manag. **39**(1), 1–14 (2001)

Chuttur, M.: Overview of the technology acceptance model: origins, developments and future directions. Work. Pap. Inf. Syst. **9**(37), 9–37 (2009)

Davis, F.: A technology acceptance model for empirically testing new end-user information systems: theory and results. Ph.D. Thesis. MIT Sloan School of Management (1986)

Davis, F.: Perceived usefulness, perceived ease of use, and user acceptance of information technology. MIS Q. **13**(3), 319–340 (1989)

Davis, F.: User acceptance of information technology: system characteristics, user perceptions and behavioral impacts. Int. J. Man Mach. Stud. **38**(3), 475–487 (1993)

Davis, F., Bagozzi, R., Warshaw, P.: User acceptance of computer technology: a comparison between two theoretical models. Manag. Sci. **35**(8), 982–1003 (1989)

Drucker, P.: Managing in the Next Society. Truman Talley Books, New York (2002)

Gorsuch, R.L.: Exploratory factor analysis. In: Gorsuch, R.L. (ed.) Handbook of Multivariate Experimental Psychology, pp. 231–258. Springer, New York (1998)

Grar, Y.: Il faut aller rapidement vers la généralisation du e-paiement (2016). El Watan: http://elwatan.com/economie/il-faut-aller-rapidement-vers-la-generalisation-du-e-paiement-04-10-2016-330007_111.php. Accessed 20 Dec 2016

Hair, J.F., Black, W.C., Babin, B.J., Anderson, R.E., Tatham, R.L.: Multivariate Data Analysis. Pearson – Prentice Hall, New Jersey (2006)

Heerwegh, D., Loosveldt, G.: Face-to-face versus web surveying in a high-internet-coverage population: differences in response quality. Public Opin. Q. **72**(5), 836–846 (2008)

Kahlane, A.: L'ARPT et le développement du numérique dans notre pays (2016). Le Soir d'Algerie: http://www.lesoirdalgerie.com/articles/2016/01/04/article.php?sid=189551&cid=41. Accessed 20 Dec 2016

Kotler, P.: Ten Deadly Marketing Sins: Signs and Solutions. Wiley, New York (2004)

Lamiri, M.: Utilisation des nouvelles technologies, quelques raisons du retard (2016). El Watan: http://www.elwatan.com/-00-00-0000-315975_175.php. Accessed 19 Dec 2016

Lowry, P., Gaskin, J.: Partial least squares (PLS) structural equation modeling (SEM) for building and testing behavioral causal theory: when to choose it and how to use it. IEEE Trans. Prof. Commun. **57**(2), 123–146 (2014)

Makhlouf, A., Belattaf, M.: Le commerce electronique en algerie: vers de nouvelles formes de vente en ligne. Studia Ekonomiczne **150**, 218–229 (2013)

Malhotra, N., Dash, S.: Marketing Research an Applied Orientation. Pearson Publishing, London (2011)

Roger, E.M.: The Diffusion of Innovations, 5th edn. The Free Press, New York (2003)

Rosenberg, M., Hovland, C.: Cognitive, affective, and behavioral components of attitudes. In: Attitude Organization and Change: An Analysis of Consistency Among Attitude Components, vol. 3, pp. 1–14 (1960)

Scholar Google (2016). http://scholar.google.fr/scholar. Accessed 18 Dec 2016

Thompson, B.: Exploratory and confirmatory factor analysis: understanding concepts and applications. Am. Psychol. Assoc. (2004)

Venkatesh, V.: Determinants of perceived ease of use: integrating control, intrinsic motivation, and emotion into the technology acceptance model. Inf. Syst. Res. **11**(4), 342–365 (2000)

Venkatesh, V., Bala, H.: Technology acceptance model 3 and a research agenda on interventions. Decis. Sci. **39**(2), 273–315 (2008)

Venkatesh, V., Davis, F.: A model of the antecedents of perceived ease of use: development and test. Decis. Sci. **27**(3), 451–481 (1996)

Venkatesh, V., Davis, F.: A theoretical extension of the technology acceptance model: four longitudinal field studies. Manag. Sci. **46**(2), 186–204 (2000)

Venkatesh, V., Morris, M.G., Davis, G.B., Davis, F.D.: A unified theory of acceptance and use of technology. MIS Q. **27**(3), 425–478 (2003)

Yousafzai, S., Foxall, G., Pallister, J.: Explaining internet banking behavior: theory of reasoned action, theory of planned behavior, or technology acceptance model? J. Appl. Soc. Psychol. **40**(5), 1172–1202 (2010)

Capturing Leading Factors Contributing to Consumer Engagement in Online Packaging Co-design Platform: A Focus Group Study and a Research Model Proposal

Olfa Ammar[✉] and Imen Trabelsi Trigui[✉]

Marketing Research Laboratory, Faculty of Economic Sciences and Management of Sfax,
Sfax, Tunisia
ammarolfa1@gmail.com, imentrigui@yahoo.fr

Abstract. In recent years, an important marketing trend has been developed where consumer-firm collaboration and interaction take place. Today, the value co-creation strategy is a new way that companies adopt to create value with consumers and for both of them. In areas including innovation and new product/ service development, the value co-creation remains the most appealing concept which allows the company to effectively allocate its resources in line with consumers' needs and preferences. The co-creation revolution was widely supported by the development of information and communication technologies and the web 2.0 techniques which changed consumers' role into an active partner. However, we need to know: why do consumers engage in new product development (here, olive oil packaging) through virtual interface? Based on a profound literature review related to consumer engagement in value co-creation experiences and in online environments, we arrived finally at a clear definition of consumer engagement concept that fits the context of our research. An exploratory study, employing three focus groups, was then conducted in order to explore the key factors that influence consumer engagement decision in online olive oil packaging design through a virtual co-design interface. The main findings highlight the importance of consumer intrinsic and extrinsic motivations, his innovativeness, his product involvement and his perception of the co-design platform interactivity and usability of the co-design interface as the main factors that influence consumer's engagement in the virtual co-design experience. From the exploratory study's findings, we develop a research model that addresses the antecedents of consumer engagement construct.

Keywords: Consumer engagement · Co-design · Packaging design · Value co-creation · Virtual co-design platform

1 Introduction

Since the 90s, the communication and information techniques and mass consumption revolution marked the advent of a new paradigm: the post-modernity. Within this perspective, Holbrook and Hirschman (1982) developed a new sight of consumption called 'experiential vision' which emphasis on the importance of providing consumers

© Springer International Publishing AG 2017
R. Jallouli et al. (Eds.): ICDEc 2017, LNBIP 290, pp. 64–82, 2017.
DOI: 10.1007/978-3-319-62737-3_6

with multi-sensory, cognitive and communicative experiences by engaging them in intense and memorable moments affecting their emotions and five senses (Schmitt 1999). As postmodern era has marked the evolution of new consumption trends, it has marked the emergence of a new consumer too, who 'takes functional features and benefits, product quality and a positive brand image as a given. What they want is product, communications and marketing campaigns that dazzle their senses, touch their hearts, and stimulate their minds' (Schmidt 1999, p. 57). Hence, the consumer is no longer a passive actor but a dynamic one who looks to interact with the firm and with the product in order to contribute actively in its production and/or its conception procedure.

To satisfy postmodern consumer's needs and believes, the relationship marketing (RM) drew on different marketing strategies to extend communication ways with the consumer. Especially, with the rise of digital marketing techniques, the consumer is more engaged in mutual exchange value experiences that increases his retention and enhances his satisfaction. The consumer engagement (CE) concept goes beyond the traditional relational constructs such as consumer involvement, participation and satisfaction, to outline consumer's psychological state and behavior outcomes (Brodie et al. 2001; Brodie and Hollebeek 2011). Along with this, the emergence of service dominant logic (S-D logic) (Vargo and Lusch 2004) and the value co-creation strategies, tries to make consumer-firm relationship more truthful, richer and go beyond simple transaction by providing the consumer more personalized experience to create stronger bonds (Prahalad and Ramaswamy 2004a, b).

The value co-creation can occur in a variety of contexts in particular, the new product development (NPD). Today, with the development of digital technologies and web 2.0 techniques, the consumer can easily communicate his needs and share his creative and innovative ideas (Hoyer et al. 2010). In our research, we are interested in olive oil packaging design which is often the production manager's responsibility, who generally, prefers a standardized packaging that can be easily manufactured and especially with lower costs. We aim to engage the consumer into new olive oil packaging design process as value proposition is not embedded in the product anymore but in the co-creation of interactive experiences (Vargo and Lusch 2004).

The key purpose of this paper is to develop a research framework that highlights the main antecedents of CE in such co-design experience through an online platform. Indeed, in the academic marketing literature, previous researches have well invested to define the concept of CE, delimit its theoretical foundations and identify its dimensions (Bowden 2009; Kumar et al. 2010; Brodie and Hollebeek 2011; Brodie et al. 2013). However, there still a lack of studies on CE's antecedents in accordance with consumer, product and media related factors. Indeed, before launching the co-design experience we should take in consideration and discuss the main reasons, from different sights, that influence consumer's engagement in the online co-design activity. In fact, we need to know in depth, from consumers' point of view, how they think about their engagement in such experience and the most important factors that may affect their participation decision. Hence, in addition to consumers' related factors we attempt to assess situational constraints that either restrain or encourage his engagement. From another side, previous studies on CE in value co-creation have almost completely ignored the food context and, particularly, the food packaging design. Our study attempts to fill this gap and further the researches on the co-design approach and new packaging development methods. Special interest in this research is about olive oil packaging design. Indeed, one of the most critical dilemmas of Tunisian olive oil export is the lack of added value to this flagship product. In fact, most of Tunisian olive oil exports are in bulk, and one of the promising ways to improve its value and to differentiate it from the competitors' products is to put it in suitable and attractive packaging. Certainly, the product packaging is the first step to get consumer attention and to begin the purchase process. For this reason, we thought about the co-design experience to develop new olive oil packaging that's matches consumer's preferences and respond to their needs. In summary, we aim to develop a research model that addresses the following question: What drive consumers to engage in an online olive oil packaging co-design experience?

This exploratory study expects to convey key insights for scholars and managers seeking to further their understanding of the CE concept in the context of virtual co-creation environment. First, we provide a theoretical background of CE concept, its key components and its related antecedents from the marketing literature review. Second, we explain how these can be integrated in a conceptual framework. Third, we use a qualitative study (focus groups) to evaluate the relationship between the identified concepts and to try to find other ones that have not been indicated in the literature and that the consumer consider them as critical factors that may influence his participation in the virtual co-design experience. Finally, we discuss the qualitative study's results and its limitations.

2   Theoretical Background

The engagement theory emerged with American socio-psychologists, especially with Kiesler (1971) who assumes that 'engagement' is a psychological state which reflects the relationship between the individual and his acts. In the organizational context, employee engagement has been the focus of many researches in the recent years (Rich et al. 2010; Alfes et al. 2013). With the development of information and communication techniques, the engagement theory has progressed in e-learning and distance education environments (Shneiderman et al. 1998), focusing on the importance of students' interaction with each other and with the engaging task. In addition, with the evolution of Web 2.0 techniques and the rise of social media, the engagement theory is relevant in order to examine interactions and communications between users, and, in turn, affect social media usage (Di Gangi and Wasko 2016). Thus, special attention is put on individuals' and individual/focal object (task, web interface, etc.) interactions based upon the idea of creating successful, collaborative and communicative tasks. Consequently, multiple studies have focused on the engagement concept and explained it according to specific contexts (Brodie and Hollebeek 2011; O'Brien and Toms 2010; Mollen and Wilson 2010; Van Doorn et al. 2010; Ashley and Tuten 2015; O'Brien and Cairns 2015; Storbacka et al. 2016).

In the marketing literature, many prior studies draw on the concept of consumer engagement from value co-creation (Brodie and Hollebeek 2011; Brodie et al. 2013; O'Brien and Cairns 2015; Pansari and Kumar 2016; Storbacka et al. 2016). However, there were only few studies attempted to define this concept and to distinct it from close relational concepts or to clarify its conceptual roots. Brodie and Hollebeek (2011), Brodie et al. (2013) tried to set apart the consumer engagement concept from consumer participation and consumer involvement. Indeed, they assume that the engagement concept is classified as the social, dynamic and interactive side of consumer/focal object relationship (e.g. brand, product, firm, task, etc.). From our side, we insist to distinguish the engagement construct from the commitment and the consumer experience ones as no previous researches clarified it. Indeed, in the organizational context, the employees' commitment refers to their satisfaction and identification with the organization; however, the engagement construct goes a step further and includes employee's optional and voluntary efforts to attain the organization goals (Saini 2006). In the marketing context, according to Taylor Research, if the company wants their committed customers to be indifferent toward competitors' offerings, it needs to engage them with its brand by developing exclusive offerings and opportunities to add real value and satisfy them. Hence, from the consumer perspective, we consider that deeper commitment (attitude) means higher degrees of engagement (attitude and behavior) which signify more emotions, more time, more effort and deeper and richer relationship. According to the user experience construct, engagement is a one of its outcomes that indicates its quality (O'Brien and Toms 2008). Indeed, whereas consumer experience occurs in a constant and specific moment of time, the consumer engagement construct is a continual and proactive one that arises during and after consumer experience.

The theoretical roots of CE arise from marketing relationship theory and the service-dominant (S-D) logic in the frame of interactive and value co-creation experiences. Indeed, relationship marketing highlights the importance of establishing and main-taining direct interaction between the producer and the consumer to create mutual values (Parvatiyar and Sheth 1999). Shani and Chalasani (1992, p. 44) argued that the RM is 'an integrated effort to identify, maintain and build up a network with individual consumers [...] for the mutual benefit of both sides, through interactive individualized and value-added contacts'. Seeing that the relationship marketing's goal is increasing the relational exchange with the consumer and integrating him actively in the value creation, Gruen and Hofstetter (2010) admit that this orientation extends the good-dominant perspective to the service-dominant (S-D) one where the consumer is 'always a co-creator of value' (Lusch and Vargo 2008). In fact, Vargo and Lusch (2004, p. 2) argue that: 'marketing has moved from a goods-dominant view, in which tangible output and discrete transactions were central, to a service-dominant view, in which intangi-bility, exchange processes, and relationships are central'. Thus, value is not embedded in the product anymore but in the benefit that the consumer gets out of using the product. Hence, value is highly subject to experiences (Karpen et al. 2012), where firms' resources in cooperation with consumer's ones are brought together to co-create value through interactive experiences. According to Brodie and Hollebeek (2011), the inter-active and co-creative consumers' experiences explain the act of his 'engagement' in the value co-creation process. They define CE concept as a 'psychological state that occurs by virtue of interactive, co-creative customer experiences with a focal agent/object (e.g., a brand) in focal service relationships. It occurs under a specific set of context dependent conditions generating differing CE levels; and exists as a dynamic, iterative process within service relationships that co-create value [...] It is a multidimensional concept subject to a context- and/or stakeholder-specific expression of relevant cognitive, emotional and/or behavioral dimensions' (Brodie and Hollebeek 2011, p. 260). In the following, drawing on previous researches findings, we attempt to discern CE's dimensions and antecedents in the frame of our study context.

Along with the value co-creation perspectives (virtual environments of value co-creation and social media interface, online brand community, etc.), the consumer engagement has been studied in the online user experience (O'Brien and Toms 2010; O'Brien and Carins 2015; Mollen and Wilson 2010) and gamification contexts as well (Harwood and Garry 2015). In fact, the consumer engagement could be built up through online and/or offline marketing channels and using different tactics and schemes. In the present research, the focus is on online consumer engagement in the design process of new packaging development. Therefore, it gathers the value co-creation and the online experience streams. However, in the literature, there is a rupture of evolving ideas in the context of consumer engagement. That's why, we should build on past researches (in value co-creation and online user experience) and associate the different theories associated with CE concept in order to discriminate its components and its antecedents related to our study context and objectives.

Many previous researches admitted that CE is a multidimensional concept that combines emotional (feelings), cognitive (thoughts) and behavioral (action) facets (Brodie and Hollebeek 2011; Brodie et al. 2013; Hollebeek 2011; Cheung et al. 2015; Martinez 2015). However, from the online user experience perspective, O'brien and Toms (2008) admit that the user engagement is the expression of human-computer interactivity and, unlike other engagement constructs in value co-creation context, its components integrate system and user variables (i.e. motivation, interest, aesthetic and sensory appeal, feedback, challenge). Engaging experiences draws in system feedback, novelty, challenge, interactivity, aesthetic and sensory appeal, interest, control, choice, motivation and positive affect (O'Brien 2008; O'Brien and Toms 2008). The variety of engagement dimensions (cognitive, emotions, interactivity, sensorial pleasure, physical, behavioral, social interaction, etc.) proves that they depend on the object of engagement (value co-creation, brand, social media, brand community, website, game, online platform, etc.). In an intermediate position, like the present one (value co-creation and online co-design platform), Kuvykaité and Taruté (2015, p. 655) admitted that "there is no consensus on which and what dimensions should be included in the concept of consumer engagement".

Therefore, we can see that different researches have given their own definitions and explanation of the engagement construct that fit their specific contexts. Hollebeek (2011, p. 789) acknowledged that to apply the engagement construct, we should follow the approach of 'who subject engages with what object'. In the present study, the subject is the consumer and the object is the online co-design task. Since each context determines the engagement definition, we aim to offer an acceptable definition of CE that suites the online co-design experience through an interactive interface. Based on the previous researches and the different developed definitions listed in the Table 1 below, we believe that: Consumer engagement is a multidimensional and context dependent concept composed of cognitive, affective and behavioral dimensions. It is an experiential response of consumer-virtual interface interaction and results from motivational drivers which combine consumer's and interface system's variables (e.g. motivation, interest, innovativeness, usability, aesthetic and sensory appeal, interactivity, etc.).

Table 1. Engagement definitions in previous researches

Author	Type of CE	Definition
Brodie and Hollebeek (2011)	Consumer brand engagement	A Psychological state that occurs by virtue of interactive, co-creative customer experiences with a focal agent/object (e.g., a brand) in focal service relationships. It occurs under a specific set of context dependent conditions generating differing CE levels; and exists as a dynamic, iterative process within service relationships that co-create value […] It is a multidimensional concept subject to a context- and/or stakeholder-specific expression of relevant cognitive, emotional and/or behavioral dimensions'
Hollebeek (2011)	Customer brand engagement	The level of an individual customers' motivational, brand related and context dependent state of mind characterized by specific levels of cognitive, emotional and behavioral activity in direct brand interactions
Mollen and Wilson (2010)	Online engagement	Cognitive and affective commitment to an active relationship with the brand as personified by the website or other computer mediated entities designed to communicate brand value
Porter et al. (2011)	Consumer engagement in an online social platform	A class of behaviors that reflects community members' demonstrated willingness to participate and cooperate with others
O'Brien and Toms (2008)	Online engagement	A category of user experience characterized by attributes of challenge, positive affect, endurability, aesthetic and sensory appeal, attention, feedback, variety/novelty, interactivity and perceived user control
Cheung et al. (2011)	Engagement in online social platform	The level of a customer's physical, cognitive, and emotional presence in connections with a particular online social platform

In this study, we go along Cheung et al. (2011, 2015) studies and we admit that engagement dimensions are: vigor, dedication and absorption. The vigor dimension refers to the physical aspect of the engagement and reflects the time, mental resilience and energy expended while using an online platform as a consumer. The dedication dimension refers to the emotional side of engagement and reveals the sense of enthusiasm, inspiration and challenge towards an online platform. Finally, the absorption dimension signifies the cognitive aspect of the psychological engagement and denotes consumers' fully concentration while using the online platform (Cheung et al. 2011, p. 3).

3 Engagement Platform

To reach consumer engagement in online co-creation tasks and to give them the chance to take the role of expert designer, they must find the appropriate tool to express their ideas and their preferences. Frow et al. (2015) admitted that effective co-creation depend on a platform for actors to engage. In marketing literature, the concept of engagement platforms arises from Prahalad and Ramaswamy's (2004b) and Ramaswamy and Gouillart's 2010) studies on innovation within co-creation experiences and online environments. Frow et al. (2015, pp. 472–473) classified the engagement platforms to five types (virtual and physical): (1) digital applications (e.g. web sites that extend the speed of interactions with multiple and diverse actors); (2) tools or products used in a way to connect actors (e.g. software companies providing software developer toolkits); (3) physical resources, where collaborators come together occasionally for mutual benefit, to share and enhance their knowledge (e.g., retail formats such as Apple stores); (4) joint processes involving multiple actors (e.g., P&G's 'connect+develop' innovation initiative); and (5) dedicated personnel groups (e.g. call center teams). An organization can combine those different types of platforms or use them separately.

In this study, we focus on online engagement platforms where the consumer can participate in the design process of new olive oil packaging. This co-design interface is an interactive and a participatory virtual workplace including co-designing tools. Certainly, this new way of new product development changes the landscape of the design process. Precisely, it modifies what we design, how we design it and who design it.

4 Research Model and Hypotheses

In this study, we develop a conceptual understanding of consumer engagement in new packaging co-design task. To reach our aim, we need to identify the major categories of antecedent variables. In their research on customer engagement behavior, Van Doorn et al. (2010) proposed a conceptual framework involving the leading key types of antecedent of customer engagement: customer-based (e.g. perceived benefits/costs, satisfaction, resources, identity, etc.), the context-based (e.g. competitive factors, social context, technological context, etc.) and the firm-based (e.g. reputation, brand characteristics, etc.) factors that can directly affect CE. In this paper, we go along Van Doorn et al. (2010) scheme, however, our main focus is on consumers' attitudes, psychological factors and perceptions, as well as, on the context-based factors that may affect CE decision and we exclude the firm-based ones.

First and foremost, we need to understand the broader question of why individuals engage in new product development experience. Certainly, the engagement decision is affected by one's specific personality and motives. Hence, identifying the motivating factors is crucial from consumers' interest about co-design tasks viewpoint, as well as, for managers in order to know how to design such virtual interfaces in the way to be extremely attractive and appealing to potential participators (Nambisan and Baron 2009). Since the motivations sustain consumers' positive attitude toward the co-design experience, in this study we focus on consumer motivations that lead to engage in the co-design task.

In consumer psychology literature, there are different theories that explain the motivational factors leading the individual to choose an alternative to the detriment of other alternatives. Certainly, consumer engages in different activities depends on his motivational state, his need to receive and his capacity to give (Hirschman 1987). This is in line with the social exchange theory (Emerson 1981) which assumes that the key force in interpersonal relationships is the satisfaction of both exchange parties by using cost-benefit comparative alternatives (Bruhn 2003). Based on this reasoning, Hoyer et al. (2010) grouped consumer motivators to psychological factors, social factors, technological factors and financial factors. However, Füller (2006) rethink this theory and admit that even if it describes the basic conditions under which consumers engage in value co-creation activity, it doesn't offer a clear insight about the endogenous variable that may explain why consumers do engage in virtual new product development task. The author believes that "consumers may engage in virtual interaction because they show a certain interest for it" (Füller 2006, p. 639). In his reasoning, Füller builds on the self-determination theory (Deci and Ryan 2000) which defines the extrinsic and intrinsic sources of motivations. Indeed, Deci and Ryan (2000) acknowledge that people may be engage in different activities by interest, curiosity and other internal factors and needs inherent in human nature. Yet, the motivations are not only intrinsic, humans could be motivated by external factors such as get rewarded, social recognition and gain grades for doing the activity. Füller (2010) presents a typology of consumer motivations in the context of virtual value co-creation and classify them to intrinsic motivations (curiosity, intrinsic innovation interest, gain knowledge, dissatisfaction with the existing products) and extrinsic motivations (monetary reward and compensation). Thus, some consumers could be more willing and able to engage in co-creation either for the experience and the task itself or for the outcomes linked to their participation (Füller 2006). In Füller's (2010) study, the intrinsic and extrinsic motivations were maintained to be important factors that shape consumers' interest and expectations toward virtual co-creation activity. In this study, we are interested about consumer engagement in new olive oil packaging co-design and we'll draw on Füller's typology as it fits the context of our research and helps us responding to our research question. Hence, we believe that consumer intrinsic and extrinsic motivations significantly affect his engagement decision in the online co-design activity.

H.1: Intrinsic motivations positively affect consumers' engagement in online co-design activity.
H.2: Extrinsic motivations positively affect consumers' engagement in online co-design activity.

However, evidence shows that consumers' motivations are not enough to generate consumer engagement in online co-design experience. The contextual factors play also an important role in shaping and enhancing the engagement decision. Thus, we should pay attention to the factors that are willing to attract or repulse consumers. This is obviously an important factor that enables an interactive experience to even have a chance oh happening. Hence, it is necessary to draw on consumers' perception of the media traits in order to start to engage them in the co-design activity.

Among context-based factors, the media interactivity is among the specific contextual aspects that affect consumer engagement in online context. Indeed, consumer's response to the co-design interface's structural proprieties and its virtual prototyping toolkits is his interaction with it (Mollen and Wilson 2010). Prahalad and Ramaswamy (2003), argue that customers' interactive experience and his active role in such virtual medium is of the same importance as their offline experience with the product. Indeed, interaction is an opportunity to understand, express needs, share ideas and to simultaneously assess and adapt resources (Prahalad and Ramaswamy 2004a, b). Mollen and Wilson (2010, p. 921) define perceived interactivity of an online interface as the "experiential phenomenon that occurs when a user interacts with a website or other computer mediated communication entities. Perceived interactivity is the degree to which the user perceives that the interaction or communication is two-way, controllable and responsive to their actions." Consequently, consumer's interaction in media context indicates his interaction with the media itself and with other users, through the media. Thus, the perceived interactivity is an experiential phenomenon that implicitly takes account in the cognitive processing and involvement in the activity (Mollen and Wilson 2010). Guthrie et al. (2004) admit that the engagement exceeds the task fulfillment and covers consumer's expanding efforts, being energized, being active and the complete usage of his cognitive capabilities. Hence, engagement differs from simple interaction with the co-design platform and the other participants, as it depends on the depth of participation the user is able to achieve with respect to each experiential attribute (O'Brien and Toms 2008). Consequently, Lalmas et al. (2014) admit that user engagement's measurement can only be done during or after interacting with the digital media. Consequently, Mpinganjira (2016), in accordance with Mollen and Wilson (2010) and Reitz (2007), confirms that the perceived interactivity is an antecedent of consumer engagement. From above, we believe that:

H.3. Consumer perception of the co-design platform interactivity positively affects his engagement in the co-designing experience.

In marketing literature, there is a variety of methods that studied consumer engagement construct and the different related concepts. Due to the exploratory nature of our study, we used a qualitative study to investigate the main reasons that drive consumers to engage in an online co-design experience of olive oil packaging. To conduct a qualitative study, there are different methods e.g. qualitative case study, observation, ethnography, nethnography, in depth interviews, focus groups, etc. to apply depends on the research questions and objectives (Hennink et al. 2010). For this reason we should select the most appropriate one that responds to our research objective. In this research we choose focus groups method as it helps us discuss and gain a range of views about the virtual co-design activity and to generate more insights on our research subject (Hennink et al. 2010).

5   Methodology

To answer our research question, the semi-structured questionnaire was the best method as it helps us to look into participants' thoughts, feelings and behaviors. Using a focus group method with open questions allows the respondent express freely and deeply his opinions by choosing his own words. Moreover, engaging the consumer in an online experience to design new olive oil packaging is still an ambiguous and unclear issue, especially in Tunisia. Thus, we need to explore consumer's opinions and reflections about such a virtual experience. As follows, the method we choose was the best alternative.

- Participants:

We interviewed 19 participants (6 female and 13 male) in two Tunisian cities (Sfax and Tunis). Our target population was olive oil consumers who are internet users. As most of Tunisian population, they are olive oil consumers. Thus, the first criterion for our selection was not difficult. Then, we focused on consumers who use the web daily and who are interested about web-based techniques and applications. All of them had a high school diploma: 15% are currently enrolled in master programs and 53% have master's degree and 21% have PhD degree. Most of the participants were under age 35, as well as the moderator, who drew from the interviewees the most significant thoughts and ideas.

- Questionnaire protocol and procedure:

The Appendix A presents the interview guidelines. It was designed to extract consumer's opinion about olive oil packaging design, his reaction toward a virtual co-design interface and his decision to conceive new olive oil packaging in accordance with his own preferences and consumption needs. The interviews lasted form 1 h to 1 h 30 min and the order of the questions varied sometimes depending on the conversation with the participants.

- Data analysis:

The interviews were video recorded and, then, transcribed and analyzed using qualitative software, NVivo 11. The main goal of our study is to clarify our theoretical findings related to CE in the packaging co-design experience and to explore this topic and consumers' views which may help us to gather preliminary data and to discover other constructs that might be otherwise unobservable. That's why, our coding scheme deals with consumer's opinions toward the olive oil packaging, the virtual co-design interface and their motivation and inspiration to participate in such online experience.

6   Results

Primarily, we were interested about investigating consumer's interest about olive oil in general and about its packaging in particular. After discussing their olive oil consumption behavior, participants explained their attitudes about olive oil packaging. We deduced that, even if all of the participants are olive oil consumers and consider it as a fundamental component of their diet, many of them neglect its packaging as they buy it in bulk and immediately from the manufacturer, "this is our tradition!" said one of the interviewees.

After presenting to them a bottle of extra virgin olive oil (image 1 in Appendix B) in order to reveal consumers' sensitivity toward the packaging in real time, 84% of participants have focused on the packaging design features (colors, graphics and shape), however, the rest maintained their attitude and were extremely "against olive oil packaging". Consequently, we can conclude that even though almost all the participants ignore the olive oil packaging, many of them were very attentive when evaluating the given packaging. Hence, developing a co-design platform could help to revise consumer's perception about olive oil packaging and why not drive him to buy the product whose design was jointly co-created with the firm.

The second part of our interview guidelines was about the participants' co-design knowledge and perception. We presented two photos (Appendix C) of customization platforms only to introduce to the participants the co-creation concept in general, and the co-design one in particular, and how they could use the different toolkits to design their preferred product. First, we investigate participants' point of view about this strategy and about the host company. Different opinions have been emerged. In fact, some participants supported this new "tactic" and saw it as a "useful", an "original and inspired strategy" because "no one knows consumers' preferences than the consumer himself!" In contrary, one of the participants believed that "it is marketing and promotional strategy more than a simple 'care' about consumers' preferences and needs". On the other side, there were other valuable views emphasizing on: Firstly, the importance of "consumers' competences and abilities" as a key factor for the success of such a strategy. Secondly, they put attention on participants' socioeconomic status. Indeed, one of the interviewees believed that "this kind of activity is not applicable to all countries and all consumers' cultures. Actually, in the more developed countries consumers are more interested in the packaging of the product and about participating in value co-creation tasks, while consumers in developing countries, such as Tunisia, are more interested in the product itself and they focus especially on its price and rarely on its appearance or its packaging". Along with this point of views, the discussion continued to talk over consumers' ability to help the company in the product development process. Almost, most of the participants agreed that the consumer had the proficiency to add value as "he is the one who knows perfectly the product's imperfections and deficiency". Therefore, we deduce that consumers' perception about the co-creation strategies differs from one to other as their perception about olive oil packaging design. It seems to be clear that there are some conditions related to the consumer and his personality traits that shape the co-creation activity as a new strategy that creates value for them.

Subsequently, we asked participants about their willingness to co-design new olive oil packaging by means of virtual co-design platform. Most of the responses were: "it depends on what I'll get in return". This has led us to discuss about their motivational factors, and here, most of the responses were "a monetary reward", "prize" or "a gift". One of the participants said "only if my name will be written in the new packaging". But, there were other motivations such as "curiosity about new and strange things" and the feeling about "doing something special and unusual". Hence, participants' motivations were mainly tangible and less experiential. However, some participants go much further and bring to light consumer's interest about innovation (new products, new technologies, new experiences, etc.) as an important reason that drives people to try new design tool and purchase new olive oil product. Indeed, one of the participants stated that "using an online platform to develop new products don't seduce me!." Some interviewees support the latter opinion and consider themselves to be the latest persons to try novel product/service or to engage in new experience. Contrary, other participants consider themselves as they are more likely to adopt the latest new products and to live new inspiring experiences than other people. From here, we can see clearly the emergence of consumer innovativeness concept as predictor of consumer behavior and engagement decision. However, that depends on the specific product field. Certainly, consumers who are more innovative and adopt the newest product in specific field may be straggle in another one. At that point, all the answers revealed that consumers' interest about the product type shapes his interest about the task itself and his incentives to participate in such virtual experience. Finally, arriving at the part dealing with co-design platform, the most critical point was its usability. Indeed, like any web-based application, the platform should be easy to use and to handle because the participant "will not waste his time" and "breaks his head!" That's why, the first thing that the participants will do, is to explore the interface, examine its different options and to "try it".

In summary, our interviews were marked by consumers' neglect of olive oil packaging and their indifference about participating in the virtual co-design experience. In addition, most of the participants have conditioned their engagement in such activity to the benefits they will get and to the co-design interface's usability. In the following, we discuss in depth the obtained results in order to propose, at the end, a research model that guides the design of future researches.

7 Discussion of Key Findings

The main objective of our study was to examine the key factors related to the consumer, to the product and to the co-design interface that predict consumers' engagement in the co-design experience. Indeed, this exploratory study has allowed us to identify new specific hidden information about consumer's attitudes, perceptions and motives.

From our study results, we can reinforce our hypothesis dealing with consumers' intrinsic and extrinsic motivations as predictor factors of individuals' engagement in the co-design task. On the other hand, as we mentioned above, there is emergence of new concept that seems to affect consumer engagement decision. It is consumer innovativeness, which indicates, according to Hirschman (1980), consumer novelty seeking and his desire to obtain information about innovations. In marketing literature, there was a debate that consumer innovativeness determines new-product adoption behavior (Goldsmith et al. 1995; Manning et al. 1995). In the context of our study, the co-design platform is a new issue for the consumer from which he can develop new olive oil packaging according to his preferences and his needs. Consequently, consumer innovativeness may affect his engagement decision in the co-design experience. Hence, we propose that:

P.1: Consumer innovativeness positively affects his engagement in the co-design activity.

Furthermore, our findings highlight the importance of consumer product involvement in shaping his disposition to engage in the co-design activity. Indeed, many participants highlighted the importance of being interested about a particular product as it manipulates their motivation to participate in such virtual experience. In addition, in marketing literature and from the communication perspective, consumer's involvement levels in a product category affect his decision process to be active or passive with firm's communication. Indeed, more highly involved consumers evaluate message information more carefully and rely on the message to form their attitudes (Silayoi and Speece 2004). From packaging perspective, if the product does not stimulate consumer's interest, he will not give much attention to it (Hughes et al. 1998). Certainly, high involvement reflects more personal relevance and importance toward a product category. Thus, being high involved activates powerful feelings of passion in people. As follows, they are more motivated to interact with others who share the same concerns (Kim and Jin 2006). Thus, besides being motivated, product involvement determines if the co-design experience is effectively relevant to the consumer to engage in. In this study we consider that consumer involvement moderate the relation between consumer motivation to co-design new olive oil packaging and his engagement in the co-design task. Indeed, the level of internal force that pushes the consumer to engage in the co-design task is moderated by his level of involvement in a product category.

P.2a: Consumer involvement in the product category moderates the impact of his extrinsic motivations on his engagement in the co-design activity.
P.2b: Consumer involvement in the product category moderates the impact of his intrinsic motivations on his engagement in the co-design activity.

Additionally, the interviewees highlighted the important role of the platform usability in facilitating their interaction with it and, then, their engagement in the co-design task. Certainly, the interface usability is an important basic of user experience. Pearson et al. (2007) define perceived usability as "the degree to which the organization of the website allows the user to perform a task quite easily and fairly quickly". Indeed, the ease of use and the navigation are the critical components of a website. Van der Heijden (2003) admitted that perceived usability determines users' behavior and judgments for a website. Hence, consumer engagement decision is influenced by his perception of the co-design interface usability. Therefore, Seo et al. (2015) identified a negative correlation between perceived usability and user emotional engagement. However, a relation between a psychological engagement and a virtual interface perceived usability still need clarification. From above, we propose that:

P.3: Consumer perception of the co-design platform usability positively affects his engagement in the co-designing experience.

Finally, from the literature review and the focus groups analysis, we create a research model addressing the different relations explained above (Fig. 1).

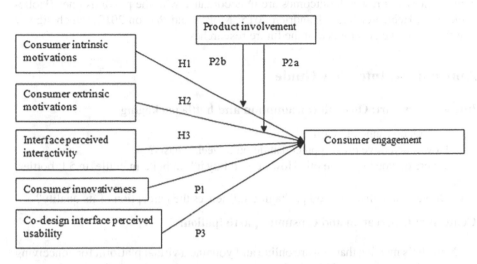

Fig. 1. Conceptual model

8   Research Implications and Limitations

This study enriches the marketing literature by providing more insights into consumer engagement concept in the context of online experience via virtual co-design interface in order to co-conceive new packaging design. In addition, it helps us clarify the subject and identify the important variables for our coming quantitative study. Actually, the existing studies in CE have rarely considered the online perspective and it is still not well determinate. In this article, we defined and discussed a conceptual framework that focuses on the major consumer-related and virtual co-design interface-related factors which have an effect on consumer's engagement in the co-design activity. Such relationships haven't been exposed in previous researches. Although the theoretical contributions, the key managerial implications of this research respond essentially to how mangers can improve their consumers' interactive online experience. Certainly, we defined the most important influencing factors, which is necessary for the managers to know in order to satisfy and communicate with their consumers in a way that their competitors can't provide.

From the managerial perspective, this research helps marketers, who are unfamiliar with the online co-creation activities; to be aware of the different factors that inhibit or encourage consumers' engagement. In addition, it provides new way of thinking to new product development plan and to achieve an efficient differentiation not only in the marketplace, but in consumers' mind, which is most significant.

Despite the exploratory nature of this research, several limitations have been uncovered. The current findings are founded on a specific sample of Tunisian consumers, while online experiences are often worldwide. Subsequently, with an exploratory study, we cannot make fundamental conclusions. A quantitative study based on experimental design conducted in different cultural contexts will be very relevant and valuable. Nevertheless, our research outcomes are in accordance with the previous ones (Hollebeek 2011; Heidenreich and Handrich 2015; Mollen and Wilson 2010) which signify that they could be generalized in the future researches.

Appendix A: Interview Guide

Introductory part: Olive oil consumption and bottle packaging:

1 - Do you consume olive oil?

2 - How much olive oil do you consume on a weekly basis?

3 - Where do you buy olive oil? How do you buy it? (In bulk, in bottle, in 5 L bottles, etc.)

4 - Do you think that olive oil packaging influences the perception of its quality?

Centering: Co-creation and consumer participation:

Now, let's imagine that you are online and you find a virtual platform for conceiving a new product packaging. This platform gives you a variety of tools that enable you to create a new packaging design. You can also find a forum where you can discuss and exchange information with other participants.

1 - What do you think of a company that involves consumers in the design of its products' packaging?

2 - Do you think that consumer engagement helps the company improve its products?

3 - If you have the possibility to participate in developing the packaging design of a new product using a virtual platform, would you participate?

4 - If yes, what would motivate you most?

5 - Would the product type incite you/drive you away from participating in the co-conception experience?

6 - Do you think that the platform system and variables (accessibility, ease of manipulation, design, etc.) would be determinant factors of your participation?

7 - Let's imagine that you are on the platform, what would you do, where would you start from?

Closure:

8 - Would you buy the product after participating in the design of its packaging?

Appendix B

See Fig. 2.

Fig. 2. Tunisian olive oil product

Appendix C

See Fig. 3.

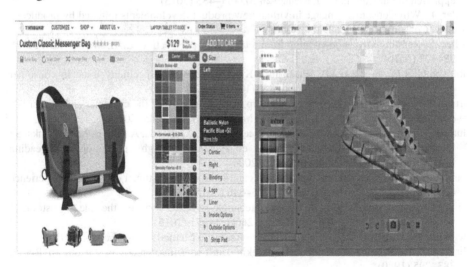

Fig. 3. Examples of customization platforms

References

Alfes, K., Shantz, A.D., Truss, C., Soane, E.C.: The link between perceived human resource management practices, engagement and employee behavior: a moderated mediation model. Int. J. Hum. Resour. Manage. **24**(2), 330–351 (2013)

Ashley, C., Tuten, T.: Creative strategies in social media marketing: an exploratory study of branded social content and consumer engagement. Psychol. Market. **32**(1), 15–27 (2015)

Bowden, J.L.: The process of customer engagement: a conceptual framework. J. Market. Theory Pract. **17**(1), 63–74 (2009)

Brodie, R.J., Hollebeek, L.D.: Advancing and consolidating knowledge about customer engagement. J. Serv. Res. **14**(3), 283–284 (2011)

Brodie, R.J., Ilic, A., Juric, B., Hollebeek, L.: Consumer engagement in a virtual brand community: an exploratory analysis. J. Bus. Res. **66**(1), 105–114 (2013)

Bruhn, M.: Relationship Marketing: Management of Customer Relationships, 289 p. Pearson Education, Harlow (2003)

Cheung, C., Lee, M., Jin, X.: Customer engagement in an online social platform: a conceptual model and scale development (2011)

Cheung, C.M., Shen, X.L., Lee, Z.W., Chan, T.K.: Promoting sales of online games through customer engagement. Electron. Commer. Res. Appl. **14**(4), 241–250 (2015)

Deci, E.L., Ryan, R.M.: Self-determination theory: a view from the hierarchical model of intrinsic and extrinsic motivation. Psychol. Inq. **11**(4), 312–318 (2000)

Di Gangi, P.M., Wasko, M.M.: Social media engagement theory: exploring the influence of user engagement on social media usage. J. Organ. End User Comput. **28**(2), 53–73 (2016)

Emerson, R.M.: Social Exchange Theory. Social Psychology: Sociological Perspectives. Basic Books, New York (1981)

Frow, P., Nenonen, S., Payne, A., Storbacka, K.: Managing co-creation design: a strategic approach to innovation. Br. J. Manag. **26**(3), 463–483 (2015)

Füller, J.: Why consumers engage in virtual new product developments initiated by producers. Adv. Consum. Res. **33**, 639–646 (2006)

Füller, J.: Refining virtual co-creation from a consumer perspective. Calif. Manag. Rev. **52**(2), 98–122 (2010)

Goldsmith, R.E., Freiden, J.B., Eastman, J.K.: The generality/specificity issue in consumer innovativeness research. Technovation **15**(10), 601–612 (1995)

Gruen, T.W., Hofstetter, J.S.: The relationship marketing view of the customer and the service dominant logic perspective. J. Bus. Mark. Manag. **4**(4), 231–245 (2010)

Guthrie, J.T., Wigfield, A., Barbosa, P., Perencevich, K.C., Taboada, A., Davis, M.H., Tonks, S.: Increasing reading comprehension and engagement through concept-oriented reading instruction. J. Educ. Psychol. **96**(3), 403 (2004)

Harwood, T., Garry, T.: An investigation into gamification as a customer engagement experience environment. J. Serv. Market. **29**(6/7), 533–546 (2015)

Heidenreich, S., Handrich, M.: Adoption of technology-based services: the role of customers' willingness to co-create. J. Serv. Manage. **26**(1), 44–71 (2015)

Hennink, M., Hutter, I., Bailey, A.: Qualitative Research Methods. Sage, London (2010)

Hirschman, E.C.: Innovativeness, novelty seeking, and consumer creativity. J. Consum. Res. **7**, 283–295 (1980)

Hirschman, E.: People as products: analysis of a complex marketing exchange. J. Market. **51**, 98–108 (1987)

Holbrook, M.B., Hirschman, E.C.: The experiential aspects of consumption: consumer fantasies, feelings, and fun. J. Consum. Res. **9**(2), 132–140 (1982)

Hollebeek, L.: Exploring customer brand engagement: definition and themes. J. Strategic Market. **19**(7), 555–573 (2011)

Hoyer, W.D., Chandy, R., Dorotic, M., Krafft, M., Singh, S.S.: Consumer cocreation in new product development. J. Serv. Res. **13**(3), 283–296 (2010)

Hughes, D., Hutchins, R., Karathanassi, V.: Purchase involvement methodology and product profiles: the case of cheese products in Greece. Br. Food J. **100**(7), 343–350 (1998)

Karpen, I.O., Bove, L.L., Lukas, B.A.: Linking service-dominant logic and strategic business practice. J. Serv. Res. **15**(1), 21–38 (2012)

Kiesler, C.A.: The psychology of commitment (1971)

Kim, H.-S., Jin, B.: Exploratory study of virtual communities of apparel retailers. J. Fashion Market. Manage. **10**(1), 41–55 (2006)

Kumar, V., Aksoy, L., Donkers, B., Venkatesan, R., Wiesel, T., Tillmanns, S.: Undervalued or overvalued customers: capturing total customer engagement value. J. Serv. Res. **13**(3), 297–310 (2010)

Kuvykaitė, R., Tarutė, A.: A critical analysis of consumer engagement dimensionality. Procedia-Social Behav. Sci. **213**, 654–658 (2015)

Lalmas, M., O'Brien, H., Yom-Tov, E.: Measuring User Engagement: Synthesis Lectures on Information Concepts, Retrieval, and Services, p. 132. Morgan & Claypool Publishers, San Rafael (2014)

Manning, K.C., Bearden, W.O., Madden, T.J.: Consumer innovativeness and adoption process. J. Consum. Psychol. **4**(4), 329–345 (1995)

Martinez, M.G.: Solver engagement in knowledge sharing in crowdsourcing communities: exploring the link to creativity. Res. Policy **44**(8), 1419–1430 (2015)

Mollen, A., Wilson, H.: Engagement, telepresence and interactivity in online consumer experience: reconciling scholastic and managerial perspectives. J. Bus. Res. **63**(9), 919–925 (2010)

Mpinganjira, M.: Influencing consumer engagement in online customer communities: the role of interactivity. Acta Commercii **16**(1), 1–10 (2016)

Nambisan, S., Baron, R.A.: Virtual customer environments: testing a model of voluntary participation in value co-creation activities. J. Prod. Innov. Manag. **26**(4), 388–406 (2009)

O'Brien, H., Cairns, P.: An empirical evaluation of the user engagement scale (UES) in online news environments. Inf. Process. Manage. **51**(4), 413–427 (2015)

O'Brien, H.L., Toms, E.G.: The development and evaluation of a survey to measure user engagement. J. Am. Soc. Inform. Sci. Technol. **61**(1), 50–69 (2010)

O'Brien, H.L., Toms, E.G.: What is user engagement? A conceptual framework for defining user engagement with technology. J. Assoc. Inform. Sci. Technol. **59**(6), 938–955 (2008)

Pansari, A., Kumar, V.: Customer engagement: the construct, antecedents, and consequences. J. Acad. Market. Sci. **45**(3), 294–311 (2016)

Parvatiyar, A., Sheth, J.N.: Handbook of Relationship Marketing, p. 680. Sage Publications, Thousand Oaks (1999)

Pearson, J.M., Pearson, A., Green, D.: Determining the importance of key criteria in web usability. Manage. Res. News **30**(11), 816–828 (2007)

Prahalad, C.K., Ramaswamy, V.: The new frontier of experience innovation. MIT Sloan Manage. Rev. **44**(4), 12–18 (2003)

Prahalad, C.K., Ramaswamy, V.: Co-creation experiences: the next practice in value creation. J. Interact. Market. **18**(3), 5–14 (2004a)

Prahalad, C.K., Ramaswamy, V.: Co-creating unique value with customers. Strategy Leadersh. **32**(3), 4–9 (2004b)

Ramaswamy, V., Gouillart, F.: The Power of Co-creation. The Free Press, New York (2010)

Rich, B.L., Lepine, J.A., Crawford, E.R.: Job engagement: antecedents and effects on job performance. Acad. Manag. J. **53**(3), 617–635 (2010)

Saini, D.S.: Managing employee relations through strategic human resource management: evidence from two Tata companies. Indian J. Ind. Relations **42**, 170–189 (2006)

Schmitt, B.: Experiential marketing. J. Market. Manage. **15**(1–3), 53–67 (1999)

Seo, K.K., Lee, S., Chung, B.D., Park, C.: Users' emotional valence, arousal, and engagement based on perceived usability and aesthetics for web sites. Int. J. Hum. Comput. Interact. **31**(1), 72–87 (2015)

Shani, D., Chalasani, S.: Exploiting niches using relationship marketing. J. Serv. Mark. **6**(4), 43–52 (1992)

Shneiderman, B., Borkowski, E.Y., Alavi, M., Norman, K.: Emergent patterns of teaching/learning in electronic classrooms. Educ. Tech. Res. Dev. **46**(4), 23–42 (1998)

Silayoi, P., Speece, M.: Packaging and purchase decisions: an exploratory study on the impact of involvement level and time pressure. Br. Food J. **106**(8), 607–628 (2004)

Storbacka, K., Brodie, R.J., Böhmann, T., Maglio, P.P., Nenonen, S.: Actor engagement as a microfoundation for value co-creation. J. Bus. Res. **69**(8), 3008–3017 (2016)

Van der Heijden, H.: Factors influencing the usage of websites: the case of a generic portal in the Netherlands. Inf. Manag. **40**, 541–549 (2003)

Van Doorn, J., Lemon, K.N., Mittal, V., Nass, S., Pick, D., Pirner, P., Verhoef, P.C.: Customer engagement behavior: theoretical foundations and research directions. J. Serv. Res. **13**(3), 253–266 (2010)

Vargo, S.L., Lusch, R.F.: Evolving to a new dominant logic for marketing. J. Market. **68**(1), 1–17 (2004)

Vargo, S.L., Lusch, R.F.: Service-dominant logic: continuing the evolution. J. Acad. Mark. Sci. **36**(1), 1–10 (2008)

Digital Economy and e-Learning

Knowledge Transfer Through E-learning: Case of Tunisian Post

Nacef Dhaouadi[✉]

Faculty of Economic Science and Management of Tunis, Tunis El Manar University,
Tunis, Tunisia
`nacef.dhaouadi@fsegt.utm.tn`

Abstract. As with ICTs today, firms are increasingly recognizing the importance of e-learning technology to train their staff and develop the necessary skills. It is therefore appropriate to consider the conditions for the success of e-learning in the organizational context, particularly from a knowledge transfer perspective. Drawing on knowledge transfer and organizational learning theories, and building on the results of an exploratory study, we propose a conceptual model that describes the determinants and consequences of e-learner satisfaction in terms of tacit knowledge transfer. An empirical investigation using the Partial Least Square (PLS) method was conducted to assess the theoretical model built.

Keywords: E-learning · E-learner satisfaction · E-learner characteristics · Organizational environment · Knowledge transfer

1 Introduction

The concept of e-learning has always interested many researchers in different fields and has been the subject of several empirical investigations until his development. The empirical studies of the e-learning have dealt with many problems due to the complexity to measure the exact impact of this technology on the various organizational and human dimensions. This complexity lies particularly in measuring the effectiveness of knowledge transfer, the success of employee ownership and the ability to replace other forms of learning (Paechter and Maier 2010). Notwithstanding, the studies carried out so far remain incomplete (Paechter and Maier 2010) and they are limited to a simple insertion of e-learning in a model of creation, transfer and knowledge dissemination, that is, a pre-established knowledge management model. Further, few studies have focused on e-learning capabilities to transfer knowledge into an organizational framework (Ou et al. 2016). This leads us to ask the following question: which organizational environment may promote the transfer of knowledge in the context of e-learning?

© Springer International Publishing AG 2017
R. Jallouli et al. (Eds.): ICDEc 2017, LNBIP 290, pp. 85–94, 2017.
DOI: 10.1007/978-3-319-62737-3_7

2   The Determinants of Knowledge Transfer Through E-learning

2.1   The Organizational Factor

The knowledge transfer throw organization can be explained by the degree of innovation in these firms (Jacob and Pariat 2000). This innovation is in fact the result of organizations adaptation to their environment changes (Cohen and Levinthal 1990) and the result knowledge transfered, shared and incorporated in organizational activities (Jacob and Pariat 2000). Moreover, if organizational learning is considered as a development of signs or information able to influence and change individual behavior (Huber 1991) organization and innovation can be also considered as the same. Starting from this idea and referring to the work of Hurley and Hult (1998), we conceive a conceptual model in the realm of "innovation market orientation & organizational learning". Hurley and Hult (1998) bring out two main categories of organizational characteristics: structure and procedure characteristics (age, degree of formalization and hierarchy) and cultural characteristics (communication, participation in decision-making, orientation towards learning). As for initiation and receptivity to innovate in organizations, they determine with good resources and other structures properties and procedures, the capacity of innovation, which is defined as the ability to adopt new ideas, procedures, or improving productivity (Fig. 1).

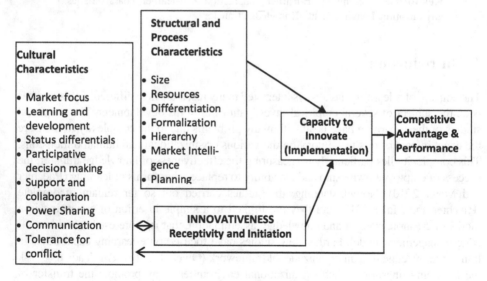

Fig. 1. Innovation, Market Orientation, and organisational learning: Learning within organization (1998)

In fact, the work of Hurley and Hult (1998) was adapted in this perspective by Graham (2003) to explain the relationship between organizational environment and knowledge outcomes. In the conceptual model of Graham (2003) individual learning is a mediator between organizational characteristics and knowledge outcomes. When it comes to corporate culture, structure and procedures, they are considered as variables likely to influence learning in organizations. Whatever the imminent or deferred use of this knowledge, this learning represents a profit for the company (Fig. 2).

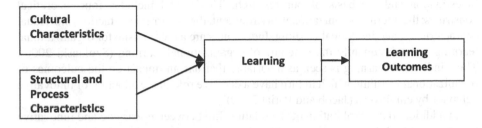

Fig. 2. Learning within organization

2.2 The Technological Factor

Theoretical Information Systems models of success have played an important role in the development of specific models for the assumption of different types of systems, such as knowledge management systems, Enterprise Resource Planning, e-Commerce, e-Government, and e-Learning (Wang 2003; Yengin et al. 2011). It should be noted that this approach involves theories in psychology and sociology in order to clarify the relationship between user behaviors, the use and user satisfaction of information technology. Thus, a link was established by this approach between the perception of users as a variable measurement and the success of setting up the information system. In this regard, Wang (2003) state that the user satisfaction of an information system is a determinant of its effectiveness. Coming to Delone and McLean (1992), they argue that the success of information system in firms is determined by six interrelated dimension. These six dimensions of success are the quality of the information, the quality of the information system, the quality of the technical service, the use of the information system, the overall satisfaction and the benefits produced by the system (Delone and McLean 1992) (Fig. 3).

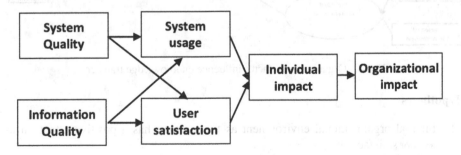

Fig. 3. Delone & McLean model

2.3   The Conceptual Model

The conceptual model that we developed represents the synthesis of our theoretical investigation and the basis of our research. This model includes four theoretical constructs: the internal organizational environment, the e-learner satisfaction, individual characteristics, and individual learning. Individuals are a prerequisite for organizational learning and subsequently the outcome of organizational learning (Szulanski 2000). Thus, individual learning is seen as a variable that has an impact on the outcomes of organizational learning, which in turn have a decisive role in the adoption of innovative behavior by employees (Jacob and Pariat 2000).

In addition to the exploration of the relationship between e-learning and innovative behavior, we will focus our study on the impact of some organizational cultural characteristics in e-learner satisfaction. The theoretical investigation has also shown that the factors of the internal organizational environment influence the thoughts and actions in the organizations, especially the organizational values that encourage learning and innovation. The organizational environment is considered in addition to e-learner satisfaction as a determinant of individual learning. Consequently, the individual learning is the expected result of the use of the e-learning system, whose organizational effects can be either imminent or potential (Fig. 4).

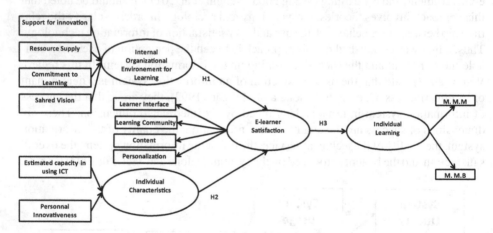

Fig. 4. Organizational factor influence on knowledge transfer

Hypotheses

H1:   Internal organizational environment as for learning has a positive effect on e-learner satisfaction.

H2:   Individual characteristics have a positive effect e-learner satisfaction.

H3:   e-learner satisfaction have a positive effect on individual learning.

3   Method

3.1   Subject and Procedure

In order to carry out our empirical study, we have administered 200 questionnaires, which represent 17% of the total number of learners enrolled in the virtual school of the Tunisian Post. This virtual school offer's a training service and professional development to the agents of more than 180 postal establishments in the world. In 2009, the virtual school of the Post Office received the award of the best distance learning content from the World Summit Award (WSA) and its platform was certified in 2012 by the Universal Postal Union (UPU). Tunisian Post has become the first partner in distance education of the UPU.

3.2   Data Analyses

Data were analyzed using SMARTPLS. PLS is a soft modeling approach to Structural Equation Method with no assumptions about data distribution (Hair et al. 2013) and less sensitive to sample size smaller than 300 (Hensler et al. 2009). Model assessment began by using confirmatory factor analysis (CFA). Construct unidimensionality, reliability and validity are tested using SPSS 18.0 and SMARTPLS 3.0. While the initial factor matrices and coefficients were calculated using SPSS, the subsequent rounds of factor analysis were performed using SMARTPLS. In fact, this is mean to validate the theoretical model by evaluating the convergence validity and the discriminant validity. The convergent validity is verified by the factor loading which must be greater than 0.70, the average variance extracted (AVE) which must be greater than 0.50, and the composite reliability of each construct which must be greater than 0.80. After having validated the model, the structural model is used to estimate the R^2 value of dependent variable. The R^2 value must be greater than 10% for any meaningful interpretation of data (Falk and miller). In addition to that, the standardized path coefficients are calculated to represent the strength of the contribution of the independent variable on the dependent variable (Chin 1998). Finally, the T statistics establish the acceptance or rejection of the study's hypotheses.

3.3   Results

The results show that the measurements are better than the cutoff value in terms of reliability. The composite reliability of the different measurements is greater than the threshold value recommended in the literature (0.707) and ranging from 0,769143 to 0,895224, indicating that the constructs are reliable and acceptable.

The results also show that the values of the square roots of the average variances extracted (AVE) of the latent variables are stronger than the correlations between these latent variable, which shows that the latent variables of the model share more variances with their indicators, than with the other variables. Cross-loadings also show that all items contribute much more to their constructs than to other constructs. Indeed, the results have revealed the presence of a value of factor weight less than 0.2 and a composite reliability equal to 0.769143 which is less than 0.8. It is worth noting that this has been verified only for a single construct, which is the support for innovation. In this regard, the measure of a formative construct is assured by examining the weight of each item, which in most cases has lower values than a reflexive construct (Karimi et al. 2007). The dimension Support for Innovation is therefore not significant, but it does contribute to the second order construct and have a T of student equal to 4.111, that's why we will not eliminate it, referring to Roberts and Thatcher (2009), who believes that conceptual is more important than statistical results ones, while you decide on the elimination of the formative indicator. Furthermore, the results show that there's no problem of multicollinearity and all the values of the VIF are greater than 1 and less than 10. Similarly, the values of R^2 mentioned in the following table, after a Bootstrap simulation, show that values are greater than 0.19, as required by Chin (1998). In this respect, the R^2 of the global model amounts to 0.460 and this is establishing good criteria for a meaningful interpretation of the results. Finally, in order to test the hypotheses of the model, we have checked the structural coefficients which, according to the results shown in the following table, show that they are a significant and positive signs for all the structural relations. Therefore, all hypotheses tested at the structural level are positively validated.

4 Discussion

4.1 Impact of the Internal Organizational Environment on the E-learner Satisfaction

The results of the empirical research assert that the organizational environment is a significant predecessor to the e-learner satisfaction, which is in line with what has been advanced in the literature. This indicates that the organizational climate geared on learning and innovation is determinant towards the involvement of the employee in the use of e-learning technologies. This observation is going along with the "planned behavior" theory of Ajzen (1991) which postulates that subjective norms such as resources and opportunities are important predecessors of intention and behavior. In fact, the organizational environment plays an important role in promoting a set of shared values among employees that adopt the use of technologies for learning (Lipshitz et al. 1996). For Degnan and Petersen (1999), the creation of a favorable climate for the use of learning technologies is the most difficult stage in the implementation of these technologies. Thereby, interest is focused especially on factors that can promote the use of e-learning technologies. The use of e-learning technologies is not only determined by its own technical characteristics, but by the synergy created between technology and the organizational environment (Guiderdoni-Jourdain 2009).

4.2 The Impact of Individual Characteristics on the E-learner Satisfaction

The results show that the e-learner satisfaction somehow depends on users' abilities to explore and exploit them, to be satisfied with what they offer as opportunities of learning outcomes. Indeed, many studies have linked individual characteristics with e-learner satisfaction (Piccoli et al. 2001; Hong 2002; Sun et al. 2008; Sawang et al. 2013). A common individual characteristic found across studies relates to learners' self-efficacy (Sawang et al. 2013) what is derived from Bandura's (1982) social learning theory, and which explains that efficacy expectations can affect intrinsic motivation for performing a task. In this respect, DeSanctis and Poole (1994) suggest that individuals have a greater effect on the use of technology in an organizational environment than the technology itself. That is to say, the characteristics of individuals have a greater impact on ICT use and satisfaction than the technical performance of these technologies. According to this, several e-learning studies have shown that learners with better computer skills have reported higher levels of satisfaction with e-learning technology (Hong 2002). In addition Sawang et al. (2013) found that in e-learning context, individuals who were openness to change were also more willing to try new ideas and are expected to be more likely to adopt new e-learning strategies as a part of their learning and development.

4.3 The Impact of E-learner Satisfaction on the Individual Learning

The results of the empirical research have demonstrated a positive and significant relationship between e-learner satisfaction and individual learning, as indicated in H3 hypothesis. This is in line with previous research specially Kane and Alavi (2007) who's already argued that the use of ICTs can enable, support and even improve the creation, the storage, the transfer and the application of knowledge in organizations (Kane and Alavi 2007). This is confirmed too by Sousa and Pinto (2013), how think that e-learning "allows learning, by relating new knowledge with past experiences, trough the linking of learning to needs, and then by practically applying the learning. This can potentially develop a more user oriented and effective deployment of knowledge availability and the development experiences". Ou et al. (2016) have found too that the effective use of systems as e-learning technology what is equivalent to the user satisfaction in our study, can significantly improve individual work performance and impart positive effects to collective network efficacy.

5 Contributions

The first managerial contribution of this work lies in the confirmation of the ability of e-learning technology to transfer all types of knowledge within organizations and thus contribute to organizational learning. However, the organizational environment and individual characteristics determine the extent to which learners can explore and exploit these technologies in learning activities that support the maintenance and construction of mental models. Consequently, an organization can't expect the benefits of distance learning technologies (knowledge development, sharing and integration) unless it promotes an organizational environment that facilitates learning and innovation. The second contribution of this work is that some individual characteristics have a significant impact on the behavior of learners in relation to the learning activities offered by e-learning technology, particularly the experience in the use of learning technologies.

6 Conclusion

Knowledge is considered in this research as individual and it's acquired through an individual learning process. However, some knowledge is created through sharing common sense, and belonging to a practice community, such as the organization (Brown and Duguid 1998). Therefore, knowledge and its value are socially constructed in a particular context (Lave and Wenger 1991) and it is the groups that decides or influences its meaning and use. In this regard, e-learning technology involves organizational approval for the use of transferred knowledge in order to bring added value. In addition, the knowledge gained from these technologies is specific to the organizational context. Even though most of the e-learning technology is used individually, it remains essential to share the common sense of the learned knowledge to be applied in the organization. Thus, managers are called to facilitate the sharing of common sense by building a shared repository supported by technology and a specific organizational structure for the management of e-learning activities.

Future research on the transfer of knowledge via e-learning technology will have to specify the kind of knowledge that companies need before testing the ability of this technology in such a transfer. We also think that future researchers should study the role of tutors in the successful transfer of all types of knowledge via e-learning. Finally, except the identification of the kind of knowledge that companies are seeking to transfer and the role of tutors, we think that future research can examine the transfer of knowledge outside the intra-organizational framework, and to evaluate it as part of an inter-organizational transfer.

Annex

See Tables 1 and 2.

Table 1. Convergent validity result

	AVE	Composite reliability	Cronbachs alpha
Individual learning (APP IND)	0,517962	0,895224	0,865437
Learning community	0,640539	0,842166	0,718779
Content	0,567819	0,839789	0,745890
Personalization	0,641889	0,877481	0,813621
Commitment to learning	0,498114	0,855718	0,797227
Personal innovativeness	0,615999	0,827761	0,687997
Learner interface	0,767271	0,868302	0,696947
Resource supply	0,651326	0,848499	0,735199
Support for innovation	0,612279	0,824908	0,681237
Shared vision	0,527864	0,769143	0,549685
Estimated capacity in using ICT	0,665548	0,856179	0,751650

Table 2. Confirmatory factor analysis result

Variables	Author	N^{bre} of items	N^{bre} of items retained	% de la VE	K.M.O
Support for innovation	Scott and Bruce (1994)	7	3	52.80	0.601
Resource supply	Scott and Bruce (1994)	4	3	49.83	0.63
Commitment to learning	Sinkula et al. (1997)	7	4	45.53	0.827
Shared vision	Sinkula et al. (1997)	4	4	57.84	0.761
Learner interface	Wang (2003)	4	4	52.30	0.709
Learning community	Wang (2003)	4	4	54.89	0.682
Content	Wang (2003)	4	4	56.83	0.764
Personalization	Wang (2003)	4	4	64.19	0.782
Individual learning (APP IND)	Vandenbosch and Higgins (1996)	8	8	61.90	0.765
Personal innovativeness	Agrawal and Prasad (1998)	4	2	76.74	0.5
Estimated capacity in using ICT	Compeau and Higgins (1991)	3	3	61.90	0.765

References

Adeyinka, T., Mutula, S.: A proposed model for evaluating the success of WebCT course content management system. Comput. Hum. Behav. **26**, 1795–1805 (2010)

Agrawal, R., Prasad, J.: Are individual differences germane to the acceptance of new information technologies. Decis. Sci. **30**(2), 361–391 (1998)

Ajzen, I.: The theory of planned behavior. Organ. Behav. Hum. Decis. Process. **50**(2), 179–211 (1991)

Alavi, M., Leinder, D.E.: Knowledge management systems: conceptual foundations and research issues. MIS Q. **25**(1), 107–136 (2001)

Bandura, A.: Self-efficacy mechanism in human agency. Am. Psychol. **37**(2), 122–147 (1982)

Brown, J.S., Duguid, P.: Organizing knowledge. Calif. Manag. Rev. **40**(3), 90–111 (1998)

Chin, W.W.: The partial least squares approach to structural equation modeling. In: Marcoulides G.A. (ed.) Modern Methods for Business Research, pp. 295–336 (1998)

Cohen, W.M., Levinthal, D.: Absorptive capacity: a new perspective on learning and innovation. Adm. Sci. Q. **35**(1), 128–152 (1990)

Degnan, C., Petersen, S.: Lotus builds knowledge around R5. PCWeek **16**(12), 1–22 (1999)

Delone, W., McLean, E.R.: Information systems success: the quest for the dependent variable. Inf. Syst. Res. **3**(1), 60–95 (1992)

DeSanctis, G., Jackson, B.M.: Coordination of information technology management: team-based structure and computer-based communication systems. J. Manag. Inf. Syst. **10**(4), 85–110

Falk, R., Miller, N.: A Primer for Soft Modeling. University of Akron Press, Akron (1992)

Graham, K.: The impact of knowledge management technologies on learning within organizations: an empirical analysis. The Florida State University College of Business (2003)

Guiderdoni-Jourdain, K.: L'appropriation d'une Technologie de 1 'Information et de la Communication en entreprise à partir des relations entre Vision-Conception- Usage. Le cas d'un Intranet RH, d'un concepteur RH et de l'utilisateur Management Intermédiaire, Université de La Méditerranée, Aix-Marseille II (2009)

Hair, J.F., Hult, G.T.M., Ringle, C.M., Sarstedt, M.: A Primer on Partial Least Squares Structural Equation Modeling. Sage, Thousand Oaks (2013)

Hensler, J., Ringle, C., Sinkovics, R.: The use of partial least squares path modeling in international marketing. Adv. Int. Mark. **20**(1), 227–320 (2009)

Hong, K.S.: Relationships between students' and instructional variables with satisfaction and learning from a web-based course. Internet High. Educ. **5**(3), 267–281 (2002)

Huber, G.: Organization learning: the contribution processes and literatures. Organ. Sci. **2**(1), 88–115 (1991)

Jacob, R., Pariat, L.: Gérer les connaissances: un défi de la compétitivité du 21éme siècle (2000). www.cefrio.qc.ca

Kane, G.C., Alavi, M.: Information technology and organizational learning: an investigation of exploration and exploitation processes. Organ. Sci. **18**(5), 796–812 (2007)

Karimi, J., Somers, T.M., Bhattacherjee, A.: The impact of ERP implementation on business process outcomes: a factor-based study. J. Manag. Inf. Syst. **24**(1), 101–134 (2007)

Lave, J., Wenger, E.: Situated Learning Legitimate Perpberal Participation. Cambridge University Press, New York (1991)

Lipshitz, R., Popper, M., Oz, S.: Building learning organizations: the design and implementation of organizational learning mechanisms. J. Appl. Behav. Sci. **32**(3), 292–305 (1996)

Ou, C.X.J., Robert, M.D., Louie, H.M.W.: Using interactive systems for knowledge sharing: the impact of individual contextual preferences in China. Inf. Manag. **53**, 145–156 (2016)

Paechter, M., Maier, B.: Online or face-to-face? Students' experiences and preferences in e-learning. Internet High. Educ. **13**(4), 292–297 (2010)

Piccoli, G., Ahmad, R., Ives, B.: Web-based virtual learning environments: a research framework and a preliminary assessment of effectiveness in basic IT skills training. MIS Q. **5**(4), 401–426 (2001)

Roberts, N., Thatcher, J.B.: Conceptualizing and testing formative constructs: tutorial and annotated example. data base Adv. Inf. Syst. **3**(40), 9–39 (2009)

Sawang, S., Newton, C., Jamieson, K.: Increasing learners' satisfaction/intention to adopt more e-learning. Educ. Train. **55**(1), 83–105 (2013)

Sun, P.C., Tsai, R.J., Finger, G., Chen, Y.Y., Yeh, D.: What drives a successful e-learning? An empirical investigation of the critical factors influencing learner satisfaction. Comput. Educ. **50**(4), 1183–1202 (2008)

Szulanski, G.: The process of knowledge transfer: a diachronic analysis of stickiness. Organ. Behav. Hum. Decis. Process. **82**(1), 9–27 (2000)

Von Krogh, G.: Care in knowledge creation. Calif. Manag. Rev. **4**(3), 133–153 (1998)

Wang, Y.S.: Assessment of learner satisfaction with asynchronous electronic learning systems. Inf. Manag. **41**, 75–86 (2003)

Yengin, L., Karahoca, A., Karahoca, D.: E-learning success model for instructors' satisfactions in perspective of interaction and usability outcomes. Procedia Comput. Sci. **3**, 1396–1403 (2011)

Modeling of a Collaborative Learning Process with Business Process Model Notation

Sameh Azouzi[✉] [iD], Sonia Ayachi Ghannouchi, and Zaki Brahmi

Laboratory RIADI-GDL, ENSI, Mannouba, Tunisia
Azouzi.Sameh@springernature.com

Abstract. Distance learning is an area in progress that continues to evolve with further research. In this context, a big attention is given by researchers, developers and educators to the collaborative learning through Learning Management Systems (LMS), especially through learning environments and social networking. However, since collaboration and interaction is mostly based on text and photos, the form and effect of collaboration and interaction is relatively simple and limited. In this paper, we propose a general model for collaborative learning processes. And we focus on a part of our process, namely sub-process "produce", to show the aspect of asynchronous but also synchronous collaboration that it enables. The proposed process is modeled with the BPMN notation and implemented through Bonita's BPMS for future management of this process.

Keywords: Collaborative e-learning · BPM · BPMN · E-learning process · Web2.0

1 Introduction

Today, e-learning technology offers a wide range of new opportunities for development of education [1]. With the evolution of Web 2.0 technologies, providing rich and simple collaboration tools (wiki, blogs, social networks, forum, webinar, etc.), one of the most important benefits of e-learning is that it enables collaborative learning. In this context, Learning Management System (LMS) is the traditional approach to ensure the e-learning process. LMS usually serves as the online platform for course syllabus releasing, handouts distribution, assignments management, and course discussion to students and teachers [2].

LMSs such as Dokeos, Moodle, Claroline and Sakai have been used by numerous universities all over the world to support and improve learning of their students [3]. In fact, "the heart of Moodle is courses that contain activities and resources" (Moodle.org 2011). Currently, it is estimated that there are about 13 different kinds of activities available in Moodle 2.0.

These activities include assignments, chats, feedbacks, forums, glossaries, surveys, quizzes, wikis and workshops. Many VLEs (Virtual Learning Environment) flourish in course management and delivery through various learning activities. Activities are very significant in this context since they correspond to the ways in which online learning is managed and have contributed to learning within the VLE. However, with the new

© Springer International Publishing AG 2017
R. Jallouli et al. (Eds.): ICDEc 2017, LNBIP 290, pp. 95–104, 2017.
DOI: 10.1007/978-3-319-62737-3_8

orientations of e-learning practices, featured by social interaction, collaboration and active participation of students and teachers, the primary limitations of LMS include the lack of interaction channel and collaboration between students and educators and also the limited number of tools to track and obtain the level of collaboration and productivity for each student [2]. These limitations make LMS not sufficient for supporting collaborative e-learning in the new era of Web 2.0 which views learning as a collaborative social process [1]. On the other hand, it is currently not possible for course designers to orchestrate various educational pedagogies around these activities in an automated manner whereby course designers can gather statistical information on learning processes that could aid future pedagogical improvement. Furthermore, within the current VLEs, learning process management, re-use and collaboration is inadequate and limited.

Therefore, this article proposes and presents a new e-learning system based on BPM for Virtual Learning Environments. Its aim is to provide the functionality of management of learning process through the effective modelling of education pedagogies in the form of learning process workflows using an intuitive graphical flow diagram user interface. One of the challenges in the adoption of the BPM concept is that, since there are differences between a learning process and a business process, it is not clear whether the concept of process in BPM is compatible with the learning process. The aim is to investigate the relationships and, through implementation, to verify if it is possible to apply BPM concept to online learning process management through using a pedagogical modelling perspective.

So, according to Helic [4], collaborative learning processes are such learning processes where learning tasks are based on real-life tasks or authentic situations and typically require and motivate the co-operation or collaboration (co-construction and exchange of knowledge) of learners in a group.

Modeling has a great importance in understanding and analyzing processes. Business Process Modeling (BPM) is one of the key factors in defining service-oriented solutions for business collaborations [5]. BPMN (Business Process Modeling Notation) is designed to cover many types of modeling; we mention the collaborative diagram which depicts interactions between two or more business entities [6].

Since there is a strong similarity between a learning process and a business process in both user aspects as well as technical aspects, we believe that applying BPM to model and manage learning processes can be efficient, easy to develop and maintain, as well as powerful enough to support a wide range of common learning situations in e-learning systems [7].

In this paper, we will first present the concepts of e-learning process, Business Process Management, web2.0. Next, we will focus explain the collaborative learning. The third section is dedicated to present the collaboration and interaction in LMSs (Learning Management System). In the fourth section, we will introduce the relation between BPM and learning processes. In the next section, we will propose the collaborative learning process. Finally, a diagnosis of the evaluation process is made and some concluding remarks are given.

2 Basic Concepts

2.1 Learning Process

The learning process is a transfer of knowledge made available to the learners according to some pedagogical structure accompanied by a planning of the work that the learner must carry out in order to guarantee a good acquisition of the knowledge.

2.2 BPM (Business Process Management)

BPM is an acronym for Business Process Management. It is a key concept where business processes are put at the center of the reflection of the managers to find a culture based on the continuous improvement of the performances of the companies and where the technologies give new possibilities to a better management of the process. BPM is seen as an approach that aims to achieve a better overall view of all the company's business processes and their interactions in order to be able to optimize and automate them by using specific applications. Dominique Annet states that «By modeling processes, we aim to have an overall view of all the processes and their interactions in order» to optimize the general functioning and to automate all that can be» [8].

When we talk about BPM life cycle, the four stages of this cycle are as follows [9]:

- (Re) design and analysis of the business process: the life cycle begins with the creation of business processes either "from scratch" or by reconfiguring an existing model.
- Configuration of the information system: the business process is implemented by configuring the information system of the organization.
- Execution and supervision of the business process: in this phase, the process is executed while controlling and supervising its execution.
- Diagnosis: This phase consists of learning knowledge from the business processes in execution and using them as input for possible improvements in business processes.

2.3 Web2.0

Web2.0 is a major technology that supports dynamic and content publishing over the Internet. It involves tagging culture and uses Internet to make links and connections with information, it allows people to create, exchange, publish, and share information in a new way of communication and collaboration. Applying Web 2.0 Applications such as (Wikipedia, blogs, YouTube, social networks, RSS, webinar, tagging) to eLearning can improve interactive communication and collaboration amongst students who have similar learning resources. Another possibility is that of helping students to find the resources and share them with others in the Web-based learning. As a result, students become the consumers and producers of learning resources. Thus, Web 2.0 provides a learning environment that has the ability to change the basic nature of learning and teaching, by the development of learner controlled learning web [10].

3   Collaborative E-learning

Collaboration, primarily refers to the fact that many people interact with each other to achieve a common and centralized result. Collaborative learning is a mutual and efficient exchange between several learners in order to build and share knowledge in a community through educational strategies.

Since communication is at the heart of all forms of educational interaction, it is likely that its impact on education systems and individual teachers and learners will be significant [2]. That makes big changes in learning approaches from transmissive, behaviorist, constructivist and socio-constructivist learning models. In the context of this last approach, social networks and electronic communication technologies, with their multiple media (text, visual and voice) and their capacity to extend interaction over time and distance, are enhancing communication, cooperation and collaboration between learners and tutors [19].

Collaborative e-learning is a learning strategy which embodies the application of new technologies and where several students interact with each other in order to achieve their common goals. It uses the collaborative environment supported by the computer network to carry out the collaborative learning, in the form of group work, between teachers, tutors and students, based on their discussion, cooperation and communication [5], using the various interaction tools offered by e-learning systems such as chat, forum, wikis, social networking, etc. Indeed, collaborative e-learning refers to a second phase of e-learning based on Web 2.0 and emerging trends in e-learning. Collaborative e-learning is inspired by the popularity of Web2.0, which places increased emphasis on social learning and use of social software tools [6].

4   Collaboration and Interaction in LMSs

LMSs are designed to provide e-learning course management service to teachers, educators and administrators. They also provide some course-based collaboration and interaction service to teachers and students. By focusing on the second aspect, collaboration and interaction in LMSs is course based and the relationship between actors in a course is temporal and unequal [18] (Fig. 1).

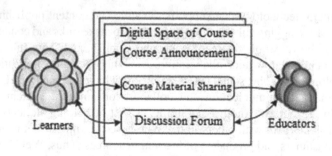

Fig. 1. Collaboration and interaction in LMS [3]

In LMSs, students and teachers can collaborate and interact within the scope of digital space of a common course. However, since collaboration and interaction is oriented to all members of a course and mostly based on text and photos, the form and effect of collaboration and interaction is relatively simple and limited. The duration of communication and interaction is the same as the corresponding course which is usually some months. Consequently, everything in the space of a course will be of no use after the end of the course, and there is no opportunity for actors (students, teachers) to accumulate personalized social network and knowledge network [3].

5 BPM and Learning Process

Business Process Management (BPM) includes methods, techniques, and software to design, enact, control, and analyze operational processes [8]. BPM is considered as an approach which aims to achieve a better overall view of business processes and their interaction in order to optimize them and automate the maximum of tasks taking advantages of devoted applications.

Helic [4] discusses the possibilities of using BPM technology for the management of collaborative learning processes. Otherwise, in e-learning scenarios, activities can be perceived as processes or workflows. A learning process represents a series of tasks or activities executed to achieve the individual or group learning goals.

Usually, in BPM a business process is defined using a graphical notation such as Petri net, YAWL (Yet Another Workflow Language), UML (Unified Modeling Language) and BPMN (Business Process Management Notation). BPMN is a comprehensible graphical notation for defining a learning process including learning activities, stakeholders, their roles and their interactions.

So, processes in e-learning context are similar to those in a business context also related to the BPM context [20].

Within this frame, several works supported the idea of modelling learning processes (collaborative) and confirm the fact that these processes can be presented by means of BPMN. According to Arnaud, the activities of learning introduce a serious need of follow-up and evaluation, which has many objectives including the fact that learners exploit at best the environment and succeed in reaching the goals of the learning activity that they realize (in terms of acquisition of knowledge, skills, etc.). Nonetheless, the domain of BP has methods and tools susceptible to serve this type of follow-up [11]. The authors in [12] suggested the BPM approach for modelling a learning process and integrated the SOA technologies to guarantee the aspect of collaboration and interaction in the learning process. Da Costa in [13] proposed a track for the development of learning devices based on scenarios by using the abstract frames of BPM. Adesina in [14] proposes an environment of virtual learning process named VLPE (Virtual Learning Process Environment) based on BPM. The idea is to propose an application "standalone" allowing modelling learning courses, which are afterward implemented via a transfer towards the BPEL execution language. Schneider in [15] modelled activities of learning with the notation BPMN.

We can conclude by considering that all of these research works concerning the systems of e-learning in the cloud and the modelling of the existing processes of e-learning suffer from some limits namely:

- The absence of a complete initiative or approach for the construction of an agile and reconfigurable process of learning adaptable to change and being able to evolve in order to meet the needs of learners;
- Lack of collaboration in learning processes and limited re-use possibilities from the processes deployed by other universities/teachers;
- The absence of an automated mechanism of communication, which informs people at the good moments when an intervention is required during the execution of the learning process.

Our objective is to adopt BPM for the construction of a general model of the process of e-learning as a collaborative, reconfigurable and executable model.

In this paper, we are interested in modeling collaborative learning process and we focus on the use of synchronous web2.0 tools such as virtual classroom, webinar, video conferencing, with the BPMN notation and thus focusing on the first phase of the BPM cycle (Design).

In fact, we focus on asynchronous and also synchronous collaborative learning processes because on the one hand, we observe that most LMSs lack synchronous collaboration. On the other hand, a BPM approach seems very useful for better management and orchestration of learning processes.

6   Collaborative Learning Process

To endorse our approach of continuous improvement regarding the learning processes based on the BPM, we propose new models of learning process based on learning activities and the interactions between learners themselves and between learners and teachers. To select the learning scenarios, we were based on Lebrun [16] and Monnard [17] models. The activities of learning are of several types: Scripting, Inform, Interact, Produce, and Formative evaluation (Fig. 2). To experience our modeling approach with BPMN, we propose to use the BPMS Bonita BPM to ensure the modeling of the collaborative learning process. This will allow to implement and manage them in future works.

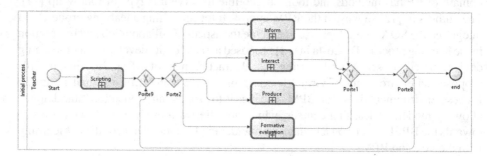

Fig. 2. Process of global e-learning

We focus through our work on the aspect of collaboration and communication between the various actors of an activity. For that purpose, and for every process, we propose sub-processes and activities of learning which run in a collaborative context with the use of tools of collaborative work such as docs.google.com, Google drive, discussion forum and webinars. As an illustration of our idea, we choose the process "Produce" which is itself detailed into two sub-processes, which are "Write an individual synthesis" and "Write a group synthesis" and trying to show the collaboration between actors in such process (Fig. 3).

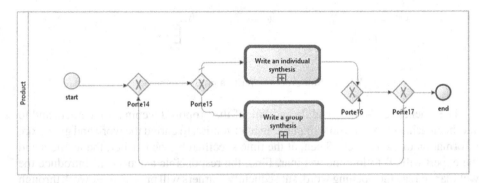

Fig. 3. Sub-process «Product»

In this paper, we will more precisely focus on the modeling of the «Write a group synthesis» process for the high level of collaboration that it involves between learner and teacher and/or between peers to perform and accomplish some activities.

In fact, collaboration between the involved actors in such a process is of two types:

- Collaboration between teacher and learner: teacher and learners may have to exchange information and documents about activities, teacher may help learners and answer their questions, as well as guide them to achieve the goals of the course.
- Collaboration between learners: learners may use web2.0 tools and create a community to prepare some activities, exchange ideas and knowledge.
- Collaboration between teachers: a teacher can give permission to another teacher or invite an expert in the field to enrich the course or perform one or many activity(ies) and thus give learners more information.

Figure 4 illustrates the process of evaluation in which the main activities included in this model are:

- The teacher adds an activity, then he/she puts the description of this activity and the link to download it through a tool from the web2.0 (on google drive for example).
- The participating learners download the activity, divide into subgroups and start the work. Then they prepare a document and share it with the teacher.
- The teacher consults the various works sent by the sub-groups and chooses the best work to make an oral presentation, then he/she launches a webinar.

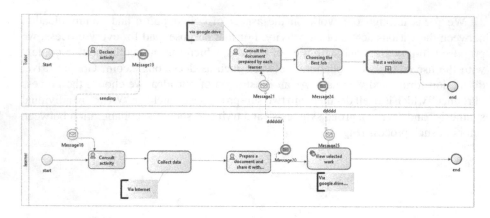

Fig. 4. Sub-process «Write a group synthesis»

The teacher sends the schedule and link of the proposed webinar to learners and to another teacher (expert) who will make an intervention to enrich the work and give more information to the learners. Then, at the time specified by the teacher, the learners and the expert will all follow the webinar. First, the responsible teacher will introduce the webinar and say the opening word, subsequently learners will present their work through presentations. When the learner concludes his/her presentation, the responsible teacher invites the expert to give comments and ends with some questions/comments and his/her own comments.

7   Evaluation of Our Process Model

Our model consists of five sub-processes as shown in Fig. 5. We choose only one part to show the synchronous and asynchronous collaboration aspect between the different players in the e-learning process (teacher, learner, expert). The objective of this work is to model a general model of e-learning process that deals with all types of activities and all forms of collaboration. Modeling is considered a fundamental first step in generating this type of process. In addition, modeling makes it possible to take into consideration all possible cases of e-learning in a collaborative context. In fact, our model can be completed and improved through the BPM lifecycle.

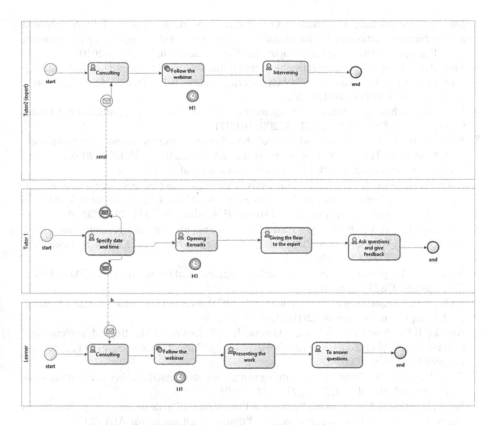

Fig. 5. Sub-process «Animate webinar»

8 Conclusion and Future Works

To conclude, the aim of this research is the development of an approach for the improvement of the process of e-learning via the Business Process Management (BPM) approach in order to make it more collaborative and general. More precisely, we presented in this paper the proposed model for our general learning process and the idea of integrating the collaborative aspect in this BPMN learning process model. In future works, we are going to investigate the possibilities offered by cloud computing technologies for a better management of our process. This will lead as to consider its a Business Process as a Service (BPaaS).

References

1. Aljenaa, E., Al-Anzi, F.S., Alshaji, M.: Towards an efficient e-learning system based on cloud computing. In: KCESS 2011 Proceedings of the Second Kuwait Conference on e-Service and e-Systems. ACM, New York (2011). ISBN: 978-1-4503-0793-2. doi: 10.1145/2107556.2107569

2. Shahid, A.N., Golam, M., Shaiful, A.C., Zakir Hossain, M., Fariha, T.J.: A Proposed architecture of cloud computing for education system in Bangladesh and the impact on current education system. IJCSNS Int. J. Comput. Sci. Netw. Secur. **10**(10), 7–18 (2010)
3. Juan, A.M., Evelio, J.G.: Implementing motivational features in reactive blended learning: application to on introductory control engineering course. IEEE Trans. Educ. **54**, 619–627 (2011). doi:10.1109/TE.2010.2102028
4. Denis, H.: Technology-supported management of collaborative learning process. Int. J. Learn. Change **1**(3) (2006). doi:10.1504/IJLC.2006.010971
5. Roser, S., Bauer, B.: A categorization of collaborative business process modeling techniques. In: E-Commerce Technology Workshops. IEEE (2005). doi:10.1109/CECW.2005.1
6. White, S.A.: Introduction to BPMN. IBM Cooperation, vol. 2(0) (2004)
7. Helic, D., Hrastnik, J., Maurer, H.: An analysis of application of business process management technology in e-learning systems. In: Proceedings of World Conference on E-learning in Corporate, Government, Healthcare, and Higher Education, pp. 2937–2942 (2005)
8. Aalst, W.M.P., Hofstede, A.H.M., Weske, M.: Business process management: a survey. In: Aalst, W.M.P., Weske, M. (eds.) BPM 2003. LNCS, vol. 2678, pp. 1–12. Springer, Heidelberg (2003). doi:10.1007/3-540-44895-0_1
9. Jean, N.G.: La gestion des processus métiers. éditeur: Lulu.com, vol. 372 (2007). ISBN: 2952826609, 9782952826600
10. Rasha, A.F.: Integration of cloud computing and web2.0 collaboration technologies in e-learning. Int. J. Comput. Trends Technol. **12**(1) (2014)
11. Angels, R.G., Miguel Angel, S.U., Garcia, E., Plazuelos, G.M.: Beyond contents and activities: specifying processes in learning technology. Current Developments in Technology-Assisted Education (2006)
12. Chua, F., Lee, C.S.: Collaborative learning using service-oriented architecture: a framework design. Knowl.-Based Syst. **22**(4), 271–274 (2009)
13. Julien, D.: BPMN 2.0 pour la modélisation et l'implémentation de dispositions pédagogiques orientées processus. Mémoire présenté pour l'obtention du master MLALT (201')
14. Ayodejil, A.: Virtual Learning process environment (VLPE): A BPM based Learning process management architecture (2013)
15. Daniel, K.: Les approches scénarisation et la modélisation du workflow pédagogique (2011)
16. Marcel, M.: La formation des enseignants aux TIC: Allier pédagogique et innovation. Institut de pédagogie universitaire et des Multimédias (IPM) (2004)
17. Gérald, C., Sergio, H., François, J., Jacque, M., Hervé, P.: Treize scénarios d'activités de cours avec Moodle. Centre NTE (2004)
18. Du, Z., Fu, X., Zhao, C., Lui, Q., Luia, T.: Interactive and collaboration e-learning platform with integrated social software and learning management system. In: Lu, W., Cai, G., Liu, W., Xing, W. (eds.) Proceedings of the International Conference on Information Technology and Software Engineering, pp. 11–18. Springer-Verlag, Berlin Heidelberg (2013). doi: 10.1007/978-3-642-34531-9_2
19. Colvin Clark, R., Richard, E.: E-learning and the science of instruction. In: Proven Guidelines for Consumers and Designers of Multimedia Learning, 4th edn. (2016)
20. Issa, T., Isaias, A.M., Kommers, P.: Multicultural awareness and technology in higher education: global perspectives. Advances in higher education and professional development (AHEPD) book series (2014)

Intermediation and Decision Support System for the Management of Unemployment: The Simulator of Duration

Anis Ben Ahmed Lachiheb[(✉)] (iD)

Economics, Management and Quantitative Finance Laboratory (LaREMFiQ),
IHEC, Sousse, Tunisia
anismax@yahoo.fr, lachiheb.anis@gmail.com

Abstract. Nowadays, many studies revealed a mismatch between job's offers and demands on job market due to several factors such as the lack of reliable data and the shortage role of public mediators. Therefore we proposed a new support system for the management of unemployment.

For this purpose, we have applied a Search Hierarchical Association Rules for Knowledge algorithm (SHARK) in order to bring light on individual determinants of unemployment duration in Tunisia. Hence, Discrete-choice models have been used to establish a accurate mechanism which describes the behaviour of long-term unemployed in Tunisia.

Thus, we developed a simulator to estimate efficiently the unemployment duration in order to enhance the process of matching by public intermediaries.

Keywords: Intermediation · Individual determinants · Unemployment duration · Discrete choice models · Algorithms · Simulator of duration

1 Introduction

The mediation on Job market is insured by a variety of private or public institutions and actors with complex and several purposes. Many empirical studies has showed the evolution of role of intermediary on the employment market and demonstrated how it's essential for the job market [1–3]. However many researches has tried unluckily to solve and find a real solution to unemployment problem [4]. Hence we are called to rebalance this hiring relationship in job market, to insure the matching by intervening, as close as possible, to the working context and much more at the heart of the professional networks [5]. Otherwise we focused on studying more closely the microeconomic individual's determinants of job seekers in order to understand and identify the issue. Unavailability of quality information is generally is fairly criticisable on several aspects specifically when provided by the public sector [6]. For some authors, individuals are unable to coordinate in the market, if it is open at fixed and regular dates [7]. The intervention of private mediators ensures accurate coordination and programmed matching. The role of the intermediary is to reduce costs and uncertainties of the

© Springer International Publishing AG 2017
R. Jallouli et al. (Eds.): ICDEc 2017, LNBIP 290, pp. 105–115, 2017.
DOI: 10.1007/978-3-319-62737-3_9

recruitment for job providers in order to guarantee the best possible matching [8–10]. IT[1] seems to appear the suitable solution to the problem [11–13].

2 Intermediation and Use of CSDM[2]S

Big Data raised as solution to technical inability to collect and use efficiently data of different types [14–16]. This confused situation motivated our choice, based on frameworks studies, to answer to this question, starting with algorithmic association's analysis [17, 18]. We noticed that few studies were interested in system interaction with human expert user (Fig. 1).

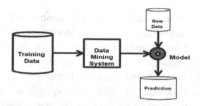

Fig. 1. Steps of the KDD process.

The beginning started with the use of large databases in commercial transactions known as Big Data [19, 20]. In our context it will be necessary to associate individual characteristics of the unemployed person to determine duration of unemployment related to them. At the same time we will be able to react on pre-established rules decision by selecting actions to reduce it, and consequently the probability of outflows of unemployment. Using KDD[3] techniques, for such massive detailed individual database, should help us to reach our target. Data mining can be defined as an iterative and interactive analysis scheme which uses raw data to extract relevant and easily and reliable information by the analyst [21]. The interactive process generated reflects the way that decision maker could analyze, control, and take corrective decisions. Then artificial intelligence methods were applied to determine the best algorithmic model. At this level, various numerical methods and visualization should enable us to evaluate effectively our issue. However it's important here to distinguish between Data Mining process (DMP)[4] and human-machine interaction (HMI). Williamson, O.E. (1991) and Elton, M.D. Book, W.J. (2010) [22, 23] has shown that DMP is based on task-oriented systems whereas HMIs are limited to define them. These specific functions represent the basis of the knowledge system. This function can be subdivided according to its specificity. The main purpose here is to realize a tool capable to estimate the duration of unemployment according to specific profile of candidate and forecast it by interacting

---

[1] Information technologies.

[2] Computer system for decision-making.

[3] Knowledge Discovery in Databases.

[4] Hegland, M. (2001). Data mining techniques. *Acta Numerica 2001, 10*, 313–355.

artificial intelligence and human touch. The major goal is to reduce the flow of jobless people and help the public intermediary to take immediate decisions to reduce their duration thanks to the use of simulator.

2.1 Empirical Approach

To establish rules on which the simulator should act to correctly determine the duration of unemployment and build recommendations, we should determine in advance a model that could forecast the individual behaviour of tunisian jobseekers according to their attributes. Our database extends over the period from January 2010 to September 2015. It's includes after deleting outliers, 206.409 males (49.7%), 208.156 females (50.3%). There are also 323.3667 unmarried (78%), 78.803 married (19%), 12.442 widowed (3%) and 37.328 divorced (9%). we proceed to a classification of individuals by date of registration, date of position, age, sex, governorate (area), delegation (district), marital status, level of education, diploma, and specialties (other details were eliminated for the purposes of the analysis and for an intelligent choice of most useful and discriminating variables[5]). Then to go further in order to associate duration to most correlated variables, the analysis through discrete choice model is highly suggested and recommended for such kind of variables [24].

Alternative of Discrete Choice Models
The evolution of econometrics has made possible to translate from an aggregated macroeconomic data analysis to a microeconomic one, thanks to the development of computerized data [25–27]. Thus, in addition to the traditional quantitative statistical analysis, appear all the interest of new treatment of qualitative variables, much more complex and most of the time neglected or omitted. Historically, the study of models describing the modalities of one or more qualitative variables began in the 1940s with relevant researches of Berkson's (1944–1951). He presented simple dichotomous models known as logit and probit models. The first attempt to apply these models in economics and political science was made by MacFadden (1974) and James J. Heckman (1976). The modelling proposed by these authors has provided a framework for the application of econometric techniques of qualitative variables for the resolution of economic problems. This has made it possible to improve the interpretation of simple models of use and information and Synthetic material (logit). It was even later developed a mid-qualitative and semi-quantitative intermediate model (Tobit model), of interest and a certain contribution, for complex and diversified problems. Therefore we proceed to the application of a multiple component analysis (MCA) to have a first idea of qualitative variables correlated to duration. Then we carry on with a multinomial logit (ML) to reveal statistical contribution of modalities of each variable (explanatory variables) on duration (explained variable). We precede further more to ordinal logit (OL) to aggregate the unemployment duration, and thus be able to compare different modelling results of qualitative variables (Fig. 2).

---

[5] According to The National Institute of statistics of Tunisia and to ILO recommendations.

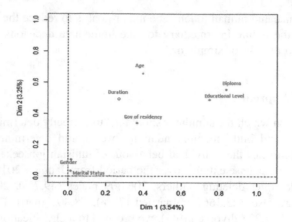

Fig. 2. The most correlated variables to the unemployment duration

MCA Analysis

The MCA model shows the correlated qualitative variables to our major explained one (correlation ratio of the variables that contribute to the dimensions of the MCA). The ultimate goal is to determine variables that are most correlated (close) to the duration (graphically). Even if this result is not very significant statistically (we talk about individual qualitative variables) it allows us to have a first idea on the evolution of the variables. A transition to logit model will be necessary to confirm or invalidate the first conclusion. We will thus focus on the representations of the active variables (formative and discriminating variables of the model) to that of the qualitative illustrative variable (the duration of unemployment on our case). Consequently:

- Gender and marital status are not graphically close to the duration, and as a consequence they do not discriminate it.
- The governorate of residence (area of belonging) and age appear to be the most relevant variables.
- Level and diploma are also related to duration, but of second importance in terms of proximity.

This figure although quite interesting, remains incomplete and should be expressed by appropriate model. The studies of Daniel L. MacFadden (1974) and James J. Heckman (1976) [28–30] suggested two kind of models which could help us: the logit or the probit. Technically the logit model is prefered because it less constraining and more general than probit modelling.

Switching to the Multinomial Logit (ML)

Modeling through multinomial logit shows that all variables explaining the duration of unemployment are significant with the exception of marital status. This model allows us to have an idea about the discriminating qualitative variables of the unemployment duration. However, going deeper on the modalities of each significant variable will be capable to generalize and aggregate the model. This empirical constraint implies a transition to ordered logit (Table 1).

Table 1. Results of modelling in multinomial logit

Variables	Criteria for fitting the model	Likelihood ratio tests		
	Log-likelihood of the model	Khi-deux	Degrees of freedom	Signif.
Constant	84344,514[a]	,000	0	.
MaritalStatu	84351,528[b]	7,014	12	,857
GouvResidence	86799,572[b]	2455,057	92	,000
Gender	87426,736[b]	3082,222	4	,000
Diploma	86815,788[b]	2471,274	16	,000
Level	85092,749[b]	748,234	16	,000
Age	90452,300[b]	6107,786	20	,000
Specialities	84468,969	124,455	28	,000
YearsDiploma	84676,538[b]	332,024	20	,000

a) This reduced model is equivalent to the final model because the omission of the effect does not increase the degrees of freedom.

b) Unexpected singularities have been found in the Hessian matrix. This indicates that some independent variables should be excluded or that some modalities should be merged.

Aggregation by Ordered Logit (OL)

Ordered logit is another model chosen by many authors of discrete choice models dealing with modelling of qualitative variables. We do first apply ordinal logit on a sample (10% of the global base) then we followed to the whole database to be able to compare and analyze the results (Table 2).

Table 2. Results of sample test in ordinal logit model

```
Iteration 0:        log likelihood                              -25942.858
Iteration 1:        log likelihood                              -25070.041
Iteration 2:        log likelihood                              -25036.199
Iteration 3:        log likelihood                              -25036.112
Iteration 4:        log likelihood                              -25036.112

Ordered logistic                        Number of obs     40679
regression                              LR ch 12(11)      1813.49
                                        Prob > chi2       0.000
Log likelihood =                        Pseudo R2         0.0358
-25036.112
```

	Coef.	Std. Err.	z	P>|z|	[95* Conf.	Interval]
Duration	.4649897	.0256818	-18.11	0.000		
sex	.018517	.0013468	13.75	0.000	-.5153251	.4146543
age	.0150822	.0346708	0.44	0.664	.0158774	.0211566
Maried	.3296339	.031245	10.55	0.000	-.0528713	.0838357
Secondary	.4270708	-.1041944	4.10	0.000	.2683949	.3908729
Secondarypr	.9734383	-.0340734	28.57	0.000	.2228535	.631288
Higher	.2467303	-.0319246	7.73	0.000	.9066556	1.040221
EastCenter	-.1452191	-.0420505	-3.45	0.001	.1841593	.3093014
westCenter	.382215	-.0460875	-8.29	0.000	-.2276366	-.0628016
Northwest	-.7520418	-.0558688	-13.46	0.000	-.4725449	-.2918851
SouthEast	.0577742	-.0486006	1.19	0.235	-.8615426	-.6425411
Southwest					-.0374812	-.1530297
/cut1	1.575093	.0654801			1.446754	1.703431
/cut2	2.375196	.0664631			2.24493	2.50546

The ordered logit on our sample determines the following aggregated model:

$$Yduration = -0.4649897Xsex + 0.018517XAge + 0.3296339XSecondary$$
$$+ 0.4270708XSecondaryprofessional + 0.9734383XHigher$$
$$+ 0.2467303XEastCenter - 0.1452191XwestCenter$$
$$- 0.382215XwestNorth - 0.7520418XEastSouth$$

As shown by the results of the model for the sample, the duration is significantly and positively correlated with all the variables of the base, except for the female gender, the belonging to the Center west, North West and South west areas, with which the duration is significant but negatively correlated. The only exception is marital status (married) which is not significant and discriminating the duration. The duration of unemployment is inversely proportional to female gender. This is a common feature for many underdeveloped countries such as ours, as reported by few publications or articles dealing with the subject [31]. The explanation of these results could be related to the low density of population there, the development of the black market with Algeria on the one hand and Libya on the other, the low rate of urbanization, the lack of employment intermediaries offices, the discouragement of young people, the low rate of direct investment, the lack of State programs and so on. We were also surprised by the fact that despite that the majority of registered unemployed are illiterates, this modality does not discriminate the duration. For level of schooling, all the modalities are significantly and positively correlated, and especially the higher level. A significant gap on the analytical spirit emerges from research on developed countries from those of less developed one on results, analyzes and reasons [32–34].

After the sample tested ordered logit, we do generalise the analysis to the entire database and then we obtain practically the same results

$$Yduration = -0.4644573Xsex + 0.0162642XAge + 0.2514821XSecondary$$
$$- 0.1187639XSecondaryProf + 0.8138134XHigher$$
$$+ 0.2625026XEastCentre - 0.1444643XWestCenter$$
$$- 0.3900611XWestNorth - 0.7119685XSouthEast + 0.06523XSouthwest$$
$$- 0.8925981XGouvernorate - 0.0389405XYearDiploma$$

Thus, to the significant variables retained by the sample, are added other significant and discriminating variables, in particular the year of graduation and especially the regional affiliation of the candidates (which is significant and negative). If we add the diploma variable, which is not significant (problem of multi-collinearity with the specialty and the level of studies) year of graduation (which is significant and negative) we obtain an aggregate ordered logit model which converges with the result of the tested sample.

2.2   Relational Schema

A decision support system (DSS) in which we have included prior probability distri-butions [35] and the SHARK (Search Hierarchic Association Rules for Knowledge) algorithm [36] is applied.

For a better interaction with the user who has the task to direct the search, an easy-to-use interface has also been developed. We used a specific graphical interface to be able to program and increment our algorithm rules relating to duration. The principle is based on the anthropocentric approach.

This lead us to reduce the expertise times and the number of association rules generated by this specific procedural method by including the user-analyst at the heart of the Data Mining process. A set of algorithms has been developed to meet these specifications. For that goal we realised is a multipurpose tool: the simulator. It first helps to calculate the average of unemployment duration for each subscripted candi-date, his probability of unemployment. It's mainly a decision- making tool able to reduce the time wasted to get a job, and consequently increasing the probability of leaving unemployment. All this required a transition to computer programming through access database to correctly define the scripts and algorithms that may establish and express relations found by ordered logit model. Thanks to this we will be able to imitate the human reasoning processes (inference, analogy and deduction) based on the available basic knowledge [37, 38].

Then we will be capable to manage individual determinants of the duration in Tunisia under an intelligent decision-making system combined with the human exper-tise granted by the agents of NEASE. A method of extraction was integrated to the system, together with validation algorithms of knowledge, connectionist and symbolic modules [39, 40]. For all that, the system was designed to generate an automatic learning for constructive acquisition of knowledge [41]. All this lead the foundation of the context in which our application over the tunisian labour market was made. The public intermediary is now given an opportunity to carry out its duties. Facing the obligation of playing perfectly its role of consultant and find ways to insure matching. The offices of this Agency have at their disposal now an instant dashboard to monitor the situation of the unemployed people and are able to advise them, and supervise the effectiveness of corrective measures of duration and hence of unemployment.

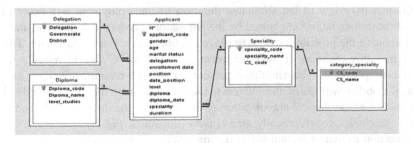

Fig. 3. Physical relational schema of the simulator

Concretely, we can increment any new registrant to automatically update the database and have accurate and unbiased calculations. This will be preceded by the integration of set of data or physical schema of the simulator (Fig. 3).

3 The Implementation of Simulator

The association of the various criteria (and/or variables) of our initial database allows us to have a double result: first, the instantaneous individual duration of each candidate according to the governorate of belonging, gender, marital status, age, diploma, and specialty. Then, the simulator determines the number of cases concerned, by such criteria in terms of population registered in the employment offices. Thus, by simulating the average duration of the population of registered unemployed male for example, without assigning precise selection criteria, we obtain an average duration of 108 days for 201,835 individuals. After that if we took randomly the governorate of Ariana as affiliation criteria, for example, in addition to male candidate and unmarried marital status, the average unemployment duration of 132.24 days is obtained, for 6407 registered unemployed (Figs. 4 and 5).

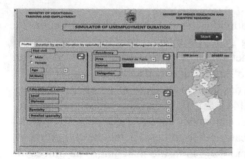

Fig. 4. Average of duration and number of cases

Fig. 5. Gender and residence affiliation

This first overview of the simulated duration of unemployment is followed by a series of recommendations suggested by the simulator to reduce this duration and to be able to improve the probability of unemployed people leaving unemployment. The aim is to reduce this duration of unemployment and thus to be able to retroact (active employment policies). The simulator presents, first, the recommendations on geographical mobility with particular interest to the areas to which the unemployed belong. An increasing chronological classification of the duration, the districts of the same area but also the closest geographically area is then illustrated (Fig. 6).

In addition to that, a second set of recommendations is proposed by the simulator according to the study specialties for the jobless whom waiting times are the longest. Their probability of leaving unemployment is then the highest. The simulator gave a

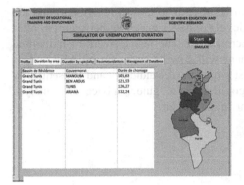
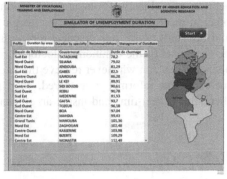

Fig. 6. Spatial recommendation of the simulator at the area and the national level

particular interest to the area to which the unemployed belong, to propose to them the district that may match the most to their profiles, their specialties and their vocational training (smallest duration) (Fig. 7).

From there, the two sets of recommendations can be combined to finally recommend the vocational training and districts, in line with the candidate's attributes or determinants. We are then facing a new tool to assist decision-maker, able to help labour market intermediaries and in particular the NEASE agency, to fulfil their role as consultant and advisor. As much as the number of criteria associated to the candidate increases, more efficient and accurate the calculations and suggested measures becomes.

Fig. 7. Chronological recommendation by speciality

4 Conclusion

This innovate computerized decision-making tool provides a source of personal and global information. This simulator determine precisely the exact duration of waiting time for a work for a particular profile. it also offers many criteria to select the weakest duration. It enables policy-makers to determine the training that generates jobs, and which should be reduced or eliminated. At the same time, this should enable us to achieve two objectives: one at the individual level (exit from unemployment) and the other at macroeconomic and even strategic level (in better matching between job offer and demand). Also, this practical tool is incremental. The management of individual determinants of the duration of unemployment will now be easier for NEASE and its agents. They can whenever retrieve the latest situation of candidates and consequently meaning the information gathered, they could inform, guide, advice and analyse thereafter at real-time and insure an efficient management.

References

1. Autor, D.H.: Wiring the labor market. J. Econ. Perspect. **15**(1), 25–40 (2001)
2. Bessy, C., Chauvin, P.M.: The power of market intermediaries: from information to valuation processes. Valuat. Stud. **1**(1), 83–117 (2013)
3. Autor, D.: The economics of labor market intermediation: an analytic framework (2008)
4. Mortensen, D.T., Pissarides, C.A.: New developments in models of search in the labor market. In: Handbook of Labor Economics, vol. 3, pp. 2567–2627 (1999)
5. Stigler, G.J.: Information in the labor market. Part 2: investment in human beings. J. Polit. Econ. **70**(5), 94–105 (1962). Published by The University of Chicago Press
6. Miller, M.H., Rock, K.: Dividend policy under asymmetric information. J. Financ. **40**(4), 1031–1051 (1985)
7. Spence, M.: Job market signaling. Q. J. Econ. **87**(3), 355–374 (1973)
8. de Larquier, G.: Principes des marchés régis par appariement. Revue économique. **48**(6), 1409–1438 (1997). Published by: Sciences Po University Press
9. Jovanovic, B.: Job matching and the theory of turnover. J. Polit. Econ. **87**(5, Part 1), 972–990 (1979)
10. Yashiv, E.: Labor search and matching in macroeconomics. Eur. Econ. Rev. **51**(8), 1859–1895 (2007)
11. Noe, R.A., Hollenbeck, J.R., Gerhart, B., Wright, P.M.: human resource management: gaining a competitive advantage (2006)
12. Bizer, C., Heese, R., Mochol, M., Oldakowski, R., Tolksdorf, R., Eckstein, R.: The impact of semantic web technologies on job recruitment processes. In: Ferstl, O.K., Sinz, E.J., Eckert, S., Isselhorst, T. (eds.) Wirtschaftsinformatik 2005, pp. 1367–1381. Physica, Heidelberg (2005). doi:10.1007/3-7908-1624-8_72
13. Rumberger, R.W., Levin, H.M.: Forecasting the impact of new technologies on the future job market. Technol. Forecast. Soc. Chang. **27**(4), 399–417 (1985)
14. Houtsma, M., Swami, A.: Set-oriented data mining in relational databases. Data Knowl. Eng. **17**(3), 245–262 (1995)
15. Kodratoff, Y.: Applications de l'apprentissage automatique et de la fouille de données. In: EGC, pp. 57–68 (2001)
16. Piateski, G., Frawley, W.: Knowledge Discovery in Databases. MIT Press, Cambridge (1991)
17. Brin, S., Motwani, R., Ullman, J.D., Tsur, S.: Dynamic itemset counting and implication rules for market basket data. ACM SIGMOD Rec. **26**(2), 255–264 (1997). ACM
18. Li, L., Zhang, M.: The strategy of mining association rule based on cloud computing. In: 2011 International Conference on Business Computing and Global Informatization (BCGIN), pp. 475–478. IEEE, July 2011

19. Leung, M.D.: Dilettante or renaissance person? How the order of job experiences affects hiring in an external labor market. Am. Sociol. Rev. **79**(1), 136–158 (2014)
20. Poterba, J.M., Summers, L.H.: Unemployment benefits and labor market transitions: a multinomial logit model with errors in classification. Rev. Econ. Stat. **77**, 207–216 (1995)
21. Williamson, O.E.: Strategizing, economizing, and economic organization. Strateg. Manag. J. **12**(S2), 75–94 (1991)
22. Elton, M.D., Book, W.J.: Operator efficiency improvements from novel human-machine interfaces. Georgia Institute of Technology (2010)
23. Takagi, H.: Interactive evolutionary computation: fusion of the capabilities of EC optimization and human evaluation. Proc. IEEE **89**(9), 1275–1296 (2001)
24. Abdi, H., Valentin, D.: Multiple correspondence analysis. In: Encyclopedia of Measurement and Statistics, pp. 651–657 (2007)
25. Pindyck, R.S., Rubinfeld, D.L.: Econometric models. In: Economic Forecasts, vol. 3 (1991)
26. McFadden, D.L.: Econometric analysis of qualitative response models. In: Handbook of Econometrics, vol. 2, pp. 1395–1457 (1984)
27. Gujarati, D.N.: Basic Econometrics. Tata McGraw-Hill Education, New York (2009)
28. Phillips, P.C., Sul, D.: Transition modeling and econometric convergence tests. Econometrica **75**(6), 1771–1855 (2007)
29. MacFadden, D.: Conditional Logit Analysis of Qualitative Choice Behavior, Major curse at University of California – Barkeley (1974)
30. Flinn, C., Heckman, J.: Models of the analysis of labor force dynamics. In: Basmann, R., Rhodes, G. (eds.) Advances in Econometrics, vol. I, pp. 35–95. JAI Press, Greenwich (1983)
31. Foley, M.C.: Determinants of unemployment duration in Russia, de l'université de Yale, Economic Growth Center, Aout 1997
32. Kahraman, E.: Youth employment and unemployment in developing country: macro challenges with micro perspectives, Ph.D., June 2011
33. McCormick, B.: A theory of signalling during job search, employment efficiency, and stigmatised jobs. Rev. Econ. Stud. **57**, 299–313 (1990)
34. Maki, D.R., Spindler, Z.A.: The effect of unemployment compensation on the rate of unemployment in Great-Britain. Oxf. Econ. Pap. **27**(3), 440–454 (1975)
35. Gregg, P., Wadsworth, J.: How effective are state employment agencies? Jobcentre use and job matching in Britain. Oxf. Bull. Econ. Stat. **58**(3), 443–467 (1996)
36. Agrawal, R., Imielinski, T., Swami, A.: Database mining: a performance perspective. IEEE Tran. Knowl. Data Eng. **5**(6), 914–925 (1993)
37. Greenberg, J., Baron, R.A.: Behavior in organizations: understanding and managing the human side of work. Pearson College Division (2003)
38. Diaper, D.: Understanding task analysis for human-computer interaction. In: The Handbook of Task Analysis for Human-Computer Interaction, pp. 5–47 (2004)
39. Han, J., Kamber, M.: Data mining: concepts and technologies. Models Methods Algorithms **5**(4), 1–18 (2001)
40. Srikant, R., Agrawal, R.: Mining sequential patterns: generalizations and performance improvements. In: Apers, P., Bouzeghoub, M., Gardarin, G. (eds.) EDBT 1996. LNCS, vol. 1057, pp. 1–17. Springer, Heidelberg (1996). doi:10.1007/BFb0014140
41. Diaper, D., Sanger, C.: Tasks for and tasks in human-computer interaction. Interact. Comput. **18**(1), 117–138 (2006)

Online Project Management and PHP7 Application: A Real Case Study

Houda Hakim Guermazi[1(⊠)] and Arij Zorai[2]

[1] LIGUE LR99ES24, Manouba University Campus, Manouba 2010, Tunisia
houda.hakim.guermazi@ensi-uma.tn
[2] National School for Computer Sciences, Manouba University Campus,
Manouba 2010, Tunisia
arij.zorai@ensi-uma.tn

Abstract. In dynamic and competitive environments, firms have several challenges to resolve in order to improve performance. However, it's difficult to manage multiple projects with multiple local or remote teammates. For this reasons, projects manager should use a tool that helps them to manage, discuss, communicate, monitor and work with team members in order to increase significantly the effectiveness of the project. Our research presents a vision of project management combining managerial and computing points of view. We present the different empirical solutions managing projects. Finally, we proposed a solution to enterprises via an online project manager developed on PHP7 focusing on three main axes: Planning, Organisation, and Control. The use of free tools minimizes the costs of integration and deployment of the proposed software. This solution allows the company to find the right balance between a good organization of the project, cost and delays.

Keywords: Online management · Project management · PHP7

1 Introduction

The concept of project management, in companies, presents an essential factor that firms can use to achieve directly or indirectly strategic objectives [38]. Eventually, in dynamic and competitive environments, project management can help to realize the success of the organization [30]. In fact, methods in project management can help to implement corporate strategies and connect the different decisions in the strategic level, operational and tactical one [7]. The project management system is a valuable tool by which firms can make their processes more organized. The authors of 29 noted that public and private companies aim and seek to perform their project management processes through maturity development management. In order to carry out project planning, companies follow certain methods that have proved their effectiveness, such as PERT, GANTT [52] and risk management [17]. However, the application of these recommendations becomes more difficult to take into account as the complexity of the project increases and as the size of enterprises increases. In this context, one of the most important improvements is to automate project management with the appropriate software to optimize firms' decision and performance in achieving objectives [16].

© Springer International Publishing AG 2017
R. Jallouli et al. (Eds.): ICDEc 2017, LNBIP 290, pp. 116–128, 2017.
DOI: 10.1007/978-3-319-62737-3_10

This research starts with a literature review of the main concepts of Project Management (PM) and online project management (Web-Based) with a focus on its importance for the enterprise. Then, we present a comparative analysis of the management project software in the international practice and we present the PHP 7. Finally, we present a real case study by introducing the problems and the adopted solution, requirements specificities and realization.

2 Literature Review of Online Project Management

2.1 Project Management

Project management (PM) become a trend known in the business world, but it has been applied successfully in different organizational and industrial sectors as diverse as the military, engineering, medicine and education technologies [19]. The technical and operational applications in the 1950s and the use of PM through the theory of contingency and the social sciences in 1970/80, led to the modern project management discipline [49]. This latter, develop rapidly through the interaction with new practices that focus on achieving better project performance by adopting complementary management practices, models and techniques from different contexts such as information management and knowledge management [28]. Then, the Project management was defined as initiating, planning, organizing, monitoring and controlling all aspects of a project, in order to achieve the objectives while respecting costs, deadlines and predefined specifications [45]. It is the application of skills, knowledge, techniques and tools to project activities to meet or exceed the expectations of the parties involved in the project [38]. The project management is constituted from five axes according to PMI (Project Management Institute) [38]: Initiating, planning, executing, controlling and closing. The most important areas in project management are: Integration, Scope, Time, Cost, Quality Procurement Human resources, Communications, Risk management and Stakeholder [20]. However, the main challenge of project management is to reach all of the project objectives [35]. The elementary constraints are scope, quality, time and budget [37]. Moreover, there are another challenge which consists on optimize and integrate the allocation of inputs [39].

2.2 Project Management Software

Project management software enables managers to plan, organize, and manage resource tools and generate practical goals and deadlines according to the data entered in the system [2]. Depending on the software, it can manage planning and estimation, scheduling, resource allocation, cost control and budget management, collaboration software, communication in order to ensure complementarities team members, documentation or administration systems. In addition, the software can offer visual ticklers tool such as Gantt Charts [52]. Today, numerous desktop and browser project management software solutions exist, and are finding applications in almost every type of business [2]. In the literatures, the important types of project management software are desktop application or Web-based (a web application used online).

2.3   The Importance of Online Project Management (Web-Based)

Online Project management is a web application, in which the user has an access using a web browser, a Smartphone or a tablet. Online Project Management is one the web application, presenting a common model for many business applications. The Online Project Management presents advantages such as the ease of accessibility from any computer connected to the internet without installing software and the collaborative work.

In any firm nowadays, project management is a very crucial aspect to achieve operation's objective [21]. To achieve the expected success, companies should identify and satisfy client needs more effectively than their competitors [5]. Practitioners and researchers see client integration as a total priority for project-based enterprises [3, 51] and a vital success factor for the project [36]. In addition, according to researchers, companies need to focus on managing project stakeholders [1, 4, 12, 54] and knowledge management [22, 42, 50] for better project quality. For these reasons, the analysis of versatile project management software show that they allow the company to add users, develop projects and activities and streamline its various operations, while reducing company's expenses and minimizing the time to do it. Hence, if the company finds a platform that can be integrated with the rest of the business operations, then it finds a solution that is worth buying.

2.4   Comparative Analysis of Existing Methods of Online Project Management in Literature Revue and International Practices

The literature revue and the study of the international practices present more than 160 software managing project. Since we cannot study and compare all software (this is not the objective of our article) we have chosen those that are most used on the market. We present, in the following Tables (1, 2, and 3) key considerations and comparison features:

Table 1. Comparative table of PM software based on general information

Software	SaaS[1]	License	Support	Score[2]	Free
Wirke	Yes	Proprietary	Email	9.7	No
eXo Platform	Yes	Proprietary, Open Source	Live support	9.4	No
JIRA	Yes	Proprietary	Email	9.4	No
Projectplace	Yes	Proprietary	Live support	9.4	No
Mavenlink	Yes	Proprietary	Training	9.3	No
Workfront	Yes	Proprietary	Phone	9.2	No
Clarizen	Yes	Proprietary	Phone	9.5	No

[1]SaaS: Software as a Service, proposes to consume software in the form of a hosted service. SaaS-based solutions cover the major functional domains: ERP, CRM, Analytics, logistics, business applications…

[2]Score rank given by users. https://project-management-software. financesonline.com/#top10-products.

Table 2. Comparative table of PM software based on features

Software	Collaborative Software	Issue Tracking System	Scheduling	Project Portfolio	Resource Management	Document Management	Wokrflow System	Reporting and analyse
Wirke	Yes	Yes	Yes	Yes	Yes	Yes	Yes	Yes
eXo Platform	Yes	No	No	No	No	Yes	Yes	Yes
JIRA	Yes	Yes	Yes	No	No	No	Yes	Yes
Projectplace	Yes	Yes	Yes	Yes	Yes	Yes	No	No
Mavenlink	Yes	No	Yes	Yes	Yes	Yes	No	No
Workfront	Yes	Yes	Yes	Yes	Yes	Yes	Yes	Yes
Clarizen	Yes	Yes	Yes	Yes	Yes	Yes	Yes	Yes

Table 3. Comparative table of the most used PM software based on monetary features

Software	Budget Management	Time Tracking	Invoicing
Wirke	No	Yes	No
eXo Platform	Unknown	Unknown	Unknown
JIRA	No	Yes	No
Projectplace	No	No	No
Mavenlink	Yes	Yes	Yes
Workfront	Yes	Yes	Yes
Clarizen	Yes	Yes	Yes

- General Information:
- Features:
- Monetary features:

Certainly, the applications presented in the preceding tables where all of them respond to the same topics, but they differ from the point of view of the offered features.

2.5   A Comparative Analysis of Language's Tools to Develop Web Software

The study of the international practices noted that developers use PHP or java as languages to develop web software.

PHP7

PHP is a scripting language that runs on the server side [32]. PHP code is included in a standard HTML page and generates dynamic content. It is free, under free license, relatively easy to learn and use [14], multiplatform, heave to set-up, easy to integrate and does not require an application server other than a WEB server frequently Apache Server. PHP creates dynamic web pages. When the user calls the page, the contents are completely or partially generated, due to information (recovered in a form or extracted from a database). For this reason, PHP is easily maintained. PHP is best suited where a web application requires more design prospect than architectural and high business intervention security.

JAVA

Java is a computer programming language that is object-oriented, concurrent, and class-based, its syntax derives from C ++ but it has fewer low-level facilities and specifically designed to have as few implementation dependencies as possible. It is based on "write once, run anywhere" (WORA). Java applications are typically compiled to bytecode that can run on any Java virtual machine (JVM). Java is one of the most popular programming languages in use in 2016, [8, 33, 48] particularly for client-server web applications, with a reported 9 million developers.

Table 4 presents a comparative analysis of PHP7 and Java languages.

Table 4. Comparative table of PHP 7 and java

	Paradigms	Typologies	Multiplatform	Slogan	Set-up heaviness	Speed
PHP 7	OO, imperative, functional, procedural, reflexive, interpreted language	Dynamic, low	yes	PHP is a popular general-purpose scripting language that is especially suited to web development	low	Faster than java
JAVA	OO, structured, imperative	Static, strong, nominative	yes	Write once, run anywhere (WORA)	medium	

3 Real Case Study

3.1 Presentation of the Company

We realize our case study within a company created in 2011. For anonymity and security reasons, we will call it X. The company is specialized in the creation, management and referencing of websites and E-commerce website. This expertise in programming will allow it to create from scratch software or website entirely tailored to the needs of his clients and those of his partners. The company has to manage a large variety of projects with a high level of performance. To achieve this aim, we worked within a team in the company. We have designed and developed an open project manager providing to the user a valid and a functional solution at all times.

3.2 Research Methodology

The aim of the research, in an interpretative approach, is no longer to discover reality and laws but to develop an understanding of social reality [6, 47]. In this context, the researcher does not choose the companies studied, however this latter presents the problem and the researcher helps to formulate and to solve it. When the researcher tries to solve an empirical problem, he will inevitably ask theoretical questions and he will interact with his environment [13, 23, 34, 47]. The solution proposed in the case study presents a form of creation and enhancement of knowledge for companies. In a second step, the researcher will regroup the various problems that exist in practice as well as the research data in order to be able to construct part of the theories and make the appropriate tests [26].

Our research is a part of a "Research-intervention approach" [47] applying a research-intervention practice [24, 47] in which the company presented the problem and as researchers we helped to formulate and solve it [13, 18, 47]. In this paper we present only, the problem and the proposed solution for the case studied, which enriches the basis of the case studies within project management topics and Business intelligence. A team of four persons (two persons presented university researchers and two persons were engineers from the company) was selected to find solution. The study was carried out during two months in 2016 in the company. We follow the W- model of Herzlich (1993), which presents the advantage of focusing on the product risk. In a first part, we prepared mock-up and design our application for duration of a month, and in a second part, we implement our application and test it.

3.3   Requirements

After analysing the problem and the practice constraints, we proposed the following recommendations and requirements of the adopted solution:

– **Project planning:** The project manager plans the project by identifying deadlines and resources, as shown in the figure Fig. 1.

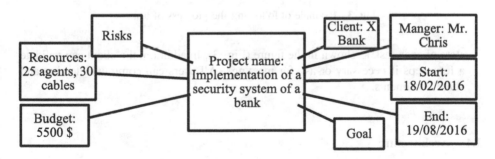

Fig. 1. Project planning explained

– **Organize tasks:** This involves distributing tasks (Fig. 2), associating them to team member according to criteria and using intelligent suggestions, identifying the relationships between them by constructing the GANTT diagram in an automatic manner and based on other diagrams such as the availability chart.
– **Control and follow the evolution of the project:** The project manager can follow the progress of the tasks and ensure the progress of the project by validating the

Fig. 2. Example of organizing the tasks of a computer project

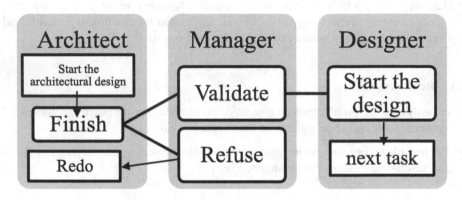

Fig. 3. Example of following the progress of tasks

state of each task as shown in the figure (Fig. 3). In addition, the client can validate a few steps if necessary or ask for meetings. All users can communicate chat and share documents.

3.4   Architecture

We choose a web application that follows 3-tier architecture as physical architecture. In addition, we followed MVC architecture Model, View and Controller as logical architecture to realise the requirements.

The view displays the data retrieved from the model and it receives all the actions of the user. The Model is responsible for retrieving the data, add, delete or modify according to the instructions sent by the controller. Finally, the controller manages query and coordinates between the model and the view.

3.5   Choice of PHP 7

In Table 4 we compared PHP7 and java, finally we have opted to use PHP7. The PHP 7 is based on PHP Data Objects (PDO) extension based on PHPNG (PHP Next Generation). It defines an excellent interface for access a database from PHP. It's faster than any other previous versions and it come to cover up the security bug. In addition, we have opted to use PHP7 because it is a major upgrade compared to the last stable version of the language, PHP 5.6. Finally, it presents an easily integration with other programs and does not require a specific editor.

3.6   Realization

In order to implement our application, we first prepared models using "Balsamiq mockups". For the graphical interfaces on the client side, we used a Bootstrap template (HTML, CSS, and JavaScript). For the server side, we used WampServer as a database server and MySQL as a database management system (DBMS) then we deployed our application under Oles Van Herman (OVH).

- **Project Planning:** In the Figs. 4 and 5 the master provides the information needed to generate a new project (project name, project manager, client and deadlines. Then displays the information of this new project and remains the database.
- **Organize tasks:** The project manager can examine the Gantt chart (Fig. 7), ensures availability of team member using the availability chart (Fig. 8). More Over, he can add a new task by providing data detailed (Fig. 6).
- **Control and follow the evolution of the project:** The project manager can also consult the progress of the tasks (Fig. 4), validate or refuse files, indicate the end of the task if successful and subsequently the end of the project.

Fig. 4. Show all tasks

Fig. 5. Add a new project

Fig. 6. Add a new task

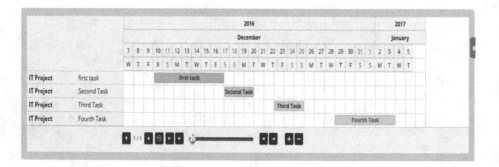

Fig. 7. GANTT

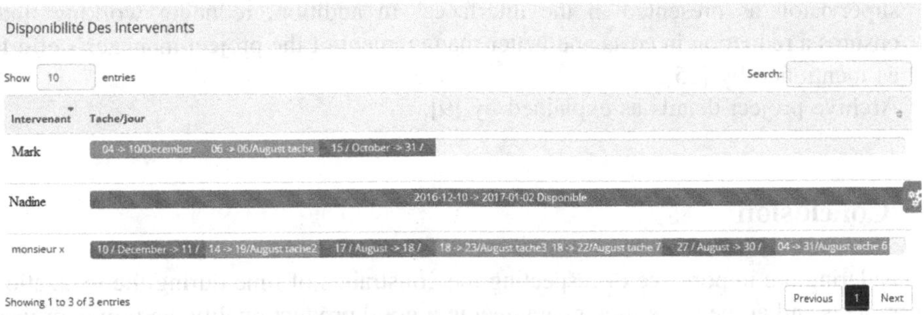

Fig. 8. Availability chart

3.7 Contribution of the Proposed Solution

Following the company's proposal to study the case, we proposed a solution for the problem which presents a form of creation and enrichment of knowledge for firms [18] and consequently an enrichment in the literature data-base dealing with a holistic vision assembling managerial and computer topics.

The literature and the international practice revue presents some common features related to the Online Project Management; we can cite creation of a workspace for the company or the work team; Creating projects, tasks and adding files; planning project and viewing files. However, these applications do not always have a complete and functional tool in all circumstances; indeed, they do not check a few options such as the verification of the validity of the dates assigned to the tasks or the availability of team members; the association of tasks according to the skills. In addition, the applications do not present a common space between all companies subscribing to the website. Finally, there is no detailed follow-up by the project manager of the evolution of the tasks of his projects. Otherwise, Applications that can fulfil the aforementioned constraints are not necessarily free and available to all users. Therefore, we proposed an application that offers to users a solution that also assign tasks according to the skills of the team members and their availability, monitoring the progress of the project by detail, and verify information before adding tasks.

Our paper presents a case study proposing a solution enabling:

- Facility of the project follow-up by using a unique code to each lot of work confirming the results of 30 dealing with rationalisation and structuring of project management activities.
- Ensure the centralization of data related to the project, thus avoiding the scattering and loss of data in the different departments related to the project. This result confirms those of [16] and those of [11].
- Improvement of the cost accuracy, time and resource estimates, which helps the project leader in decision-making and improves decision-process performance enhancing the results of [16] and those of [31].
- Help for the project manager in carrying out his/her work in the shortest possible time using GANTT diagrams in allocation of tasks, allocation of budgets, and

supervision as presented in the interfaces. In addition, reducing working time ensures a reduction in costs and better management of the project manager's efforts as mentioned by [25].
– Archive project details as explained by [9].

4 Conclusion

Considering the importance of respecting the constraints of time during the realization of projects and at the same time guaranteeing a good product quality, an online project management becomes a strategic necessity for the company which will also greatly facilitate the work for the different profiles concerned by a project namely the client, the project master and the team members. Without strong project management, the company usually duplicates its efforts, uses resources poorly, loses money due to change requests, and time as a result of poor communication. While the review of the literature has shown that many research works only deals with the general aspect of project management, and others only the existing techniques, this article try to match the two concepts of Project Management and Online Project Management using PHP7.

Moreover, we propose a solution, via a project management web application that is fully developed in PHP7 and we have specified our scope by focusing on three main axes: project planning, task organization and the follow-up of the progression. The solution assures an easy integration with the infrastructure already existent, with a high level of security. In addition, the implementation and the use of the solution offered financial and technological benefits to the studied company. This application allows the company to reduce effort and cost by proposing realistic planning and solving problems more quickly and help the project manager in the decision process. Due to the characteristics of PHP 7, this application can be expanded with new modules such as quality management and risk management.

References

1. Achterkamp, M.C., Vos, J.F.J.: Investigating the use of the stakeholder notion in project management literature, a meta-analysis. Int. J. Project Manage. **26**, 749–757 (2008)
2. Adner, R., Helfat, C.E.: Corporate effects and dynamic managerial capabilities. Strateg. Manag. J. **24**(10), 1011–1025 (2003)
3. Alajoutsijarvi, K., Mainela, T., Salminen, R., Ulkuniemi, P.: Perceived customer involvement and organizational design in project business. Scand. J. Manag. **28**, 77–89 (2012)
4. Altonen, K., Jaakko, K., Tuomas, O.: Stakeholder salience in global projects. Int. J. Proj. Manag. **26**, 509–516 (2008)
5. Artto, K., Valtakosk, A., Kärki, H.: Organizing for solutions: how project-based firms integrate project and service businesses. Ind. Mark. Manag. **45**, 70–83 (2015)
6. David, A.: Research-intervention, general framework for management research? In: David, A., Hatchuel, A., Laufer, R. (eds.) The New Foundations of Management Science, pp. 193–214. Vuibert, Paris (2000)

7. de Guimaraes, J.C.F., Severo, E.A., Vieira, P.S.: Cleaner production, project management and strategic drivers: an empirical study. J. Clean. Prod. **141**, 881–890 (2017)
8. Design Goals of the Java Programming language. http://www.oracle.com/technetwork/java/intro-141325.html. Accessed 28 Feb 2017
9. Eric McConnell: Project Records Management In Three Essential Steps (2011). http://www.mymanagementguide.com/project-records-management-in-three-essential-steps/. Accessed 02 Mar 2017
10. Ethiraj, S.E., Kale, P., Krishnan, M.S., Singh, J.V.: Where do capabilities come from and how do they matter? Strateg. Manag. J. **26**(1), 25–45 (2005)
11. Pertusa-Ortega, E.M., Zaragoza-Sáez, P., Claver-Cortés, E.: Can formalization, complexity, and centralization influence knowledge performance? J. Bus. Res. **63**(3), 310–320 (2010)
12. Freeman, E., Harrison, J., Wicks, A., Parmar, B., Colle, S.: Stakeholder Theory: The State of the Art. Cambridge University Press, New York (2008)
13. Garel, G., Giard, V., Midler, C.: Faire de la Recherche en Management de Projet. Vuibert, Paris (2004)
14. Pitkin, G.M., Ed.D.: Using Include Files to Streamline Web Site Management (2008)
15. Gosling, J., Joy, B., Steele, G., Bracha, G., Buckley, A.: The Java® Language Specification (2014)
16. Hakim Guermazi, H., Mbarek, I.: Quality management integrated system and reporting: Tunisian case study. Qual. Access Success **17**(152/June), 53–60 (2016)
17. Hakim Guermazi, H., Fourati, M.A.: Intelligent dashboard and risk management: Tunisian case study. Int. J. Bus. Econ. Strategy **4**(Special issue 1), 6 (2016)
18. Hatchuel, A.: Quel horizon pour les sciences de gestion? Vers une théorie de l'action collective. In: David, A., Hatchuel, A., Laufer, R. (eds.) Les nouvelles fondations des sciences de gestion, pp. 7–44. Vuibert, Paris (2000)
19. Ingason, H.Th., Shepherd, S.M.M.: Mapping the future for project management as a discipline - for more focused research efforts. Procedia – Soc. Behav. Sci. **119**, 288–294 (2014)
20. Ireland, L.R.: Project Management, p. 110. McGraw-Hill Professional, New York (2006)
21. Kirca, A., Jayachandran, S., Bearden, W.: Market orientation: a meta-analytic review and assessment of its antecedents and impact on performance. J. Mark. **69**, 24–41 (2005)
22. Koskinen, K., Pihlanto, P.: Knowledge Management in Project-based Companies: An Organic Perspective. Palgrave Macmillan, New York (2008)
23. Latour, B.: La Science en Action. Gallimard, Paris (1989)
24. Le Moigne J.-L.: Epistémologies constructivistes et sciences de l'organisation. In: Martinet, A.-C. (ed.) Epistémologies et Sciences de Gestion, Paris Economica, pp. 81–140 (1990)
25. Lewis, J.P.: Project Planning, Scheduling & Control, 4E. McGraw Hill, New York (2005)
26. Lundin, R.A., Wirdenius, H.: Interactive research. Scand. J. Manag. **6**(2), 125–152 (1990)
27. McMillan, R.: I Is java losing Its Mojo? (2013). https://www.wired.com/2013/01/java-no-longer-a-favorite/. Accessed 28 Feb 2017
28. Morris, P., Jeffrey, S., Pinto, J.: Towards the third wave of project management. In: The Oxford Handbook of Project Management. Oxford Handbooks Online (2011)
29. Neverauskas, B., Railaite, R.: Formation approach for project management maturity measurement. Econ. Manag. **18**(2), 360–365 (2013)
30. Niazi, M., Sajjad, M., Alshayeb, M., Rehan Riaz, M., Faisal, K., Cerpa, N., Ullah Khan, S., Richardson, I.: Challenges of project management in global software development: a client-vendor analysis. Inf. Softw. Technol. **80**, 1–19 (2016)
31. Yigitbasioglu, O.M., Velcu, O.: A review of dashboards in performance management: implications for design and research. Int. J. Acc. Inf. Syst. **13**, 41–59 (2012)

32. Olivier HEURTEL: PHP 7 - Développez un site web dynamique et interactif (2016)
33. ORACLE: Java One 2013 Review: Java Takes on the Internet of Things. http://www.oracle.com/technetwork/articles/java/afterglow2013-2030343.html. Accessed 28 Feb 2017
34. Pettigrew, A.: Longitudinal field research on change: theory and practice. Organ. Sci. **1**(3), 267–292 (1990)
35. Phillips, J.: PMP Project Management Professional Study Guide, p. 354. McGraw-Hill Professional, Emeryville (2003)
36. Pinto, J., Rouhiainen, P.: Building Customer-based Project Organizations (2001)
37. Project Management Institute: A Guide to the Project Management Body of Knowledge 4th edn., pp. 27–35 (2013)
38. Project Management Institute: A Guide to the Project Management Body of Knowledge (PMBOK Guide) (2000)
39. Project management software. https://project-management-software.financesonline.com/#history-of. Accessed 18 Dec 2016
40. Ramu, G.: Quality in outsourcing – essentials for today's global marketplace. Qual. Progress J. **22**, 37–43 (2008)
41. RedMonk Index: The RedMonk Programming Language Rankings, January 2015. Accessed 28 Feb 2017
42. Reich, B.H., Gemino, A., Sauer, C.: How knowledge management impacts performance in projects: an empirical study. Int. J. Proj. Manag. **32**, 590–602 (2014)
43. Release Candidate 6 de PHP 7.1. http://php.net/index.php#id2016-10-13-1. Accessed 20 Dec 2016
44. Sallis, E.: Total Quality Management in Education. Taylor and Francis e-Library, London (2005)
45. Sanchez, M.A.: Integrating sustainability issues into project management. J. Cleaner Prod. **96**, 319–330 (2015)
46. The complete list of developments on the PHP project website. http://php.net/index.php#id2016-07-07-1. Accessed 18 Dec 2016
47. Thiétart, R.A.: Méthodes de recherche en management. Dunod, Paris (2003)
48. TIOBE Programming Community Index. http://www.tiobe.com/tiobe-index/. Accessed 28 Feb 2017
49. Todorovic, M., Bjelica, D., Mitrovic, Z.: Concepts and models for presentation of project success. Serbian Project Management Journal Belgrade: Serbian Project Management Association- YUPMA, pp. 47–48 (2013)
50. Todorovic, M., Petrovic, D., Mihic, M., Obradovic, V., Bushuyev, S.: Project success analysis framework: a knowledge-based approach in project management. Int. J. Proj. Manag. **33**, 772–783 (2015)
51. Voss, M.: Impact of customer integration on project portfolio management and its success - Developing a conceptual framework. Int. J. Proj. Manag. **30**, 567–581 (2012)
52. Durfee, W.: Project Planning and Gantt Charts (2008)
53. Write once run anywhere. http://www.computerweekly.com/feature/Write-once-run-anywhere. Accessed 28 Feb 2017
54. Xiaojin, W., Huang, J.: The relationships between key stakeholders project performance and project success: perceptions of Chinese construction supervising engineers. Int. J. Proj. Manag. **24**, 253–260 (2006)

Data Science and Security

Data Stream Mining Based-Outlier Prediction for Cloud Computing

Imen Souiden[1](✉), Zaki Brahmi[2,3](✉), and Lamine Lafi[1,2,3]

[1] ISIGK, Kairouan University, Kairouan, Tunisia
imen.sui@gmail.com
[2] ISITcom, Sousse University, Sousse, Tunisia
zakibrahmi@gmail.com
[3] ISSAT, Sousse University, Sousse, Tunisia

Abstract. The cloud computing is the dream of computing used as utility that became true. It is currently emerging as a hot topic due to the important services it provides. Ensuring high quality services is a challenging task especially with the considerable increase of the user's requests coming continuously in real time to the data center servers and consuming its resources. Abnormal users requests may contribute to the system failure. Thus, it's crucial to detect these abnormalities for further analysis and prediction. To do that, we propose the use of the outlier detection techniques in the context of the data stream mining due to the similarity between the nature of the data streams and the users requests which require analysis and mining in real time. The main contribution of this paper consists of: first, the formulation of the users requests as well as the server state as a stream of data. This data is generated from CSG^+ a cloud stream generator that we extended from CSG [1]. Second, the creation of a framework for the detection of the abnormal users requests in terms of the CPU and memory by using AnyOut and MCOD algoithms implemented within MOA (Massive Online Analysis) (http://moa.cms. waikato.ac.nz/) framework. Third, the comparison between them in this context.

Keywords: Data stream mining · Outlier detection · Cloud computing

1 Introduction

Due to the emergence of the new technologies, such as the virtualization, autonomic computing, grid computing and utility-based pricing in addition to the success of Internet, the cloud computing has appeared as a new paradigm that reshapes the IT services provided to users. It gives new opportunities and increases the ease of use by providing computing and storage resources as services via Internet. These services are delivered on demand, with a pay per use fashion. They are hosted, deployed and stored within the virtual machines residing in the data centers servers. Therefore, users become free from many constrains that bound their usage. This encourages them to produce, consume and share more

© Springer International Publishing AG 2017
R. Jallouli et al. (Eds.): ICDEc 2017, LNBIP 290, pp. 131–142, 2017.
DOI: 10.1007/978-3-319-62737-3_11

data continuously, in real time. Thus, nowadays cloud computing has a primordial importance. The failure and the damage in data centers providing services may contributes to many critical problems. Therefore, it's crucial to ensure the consistency of the service provisioning. Mainly, this failure may be a reason of malicious behaviors (virus, spam...), panes or errors residing in the servers holding virtual machines that execute the users' requests (tasks). Thus, an analysis of data coming to cloud data centers servers in forms of users requests must be elaborated, and the abnormal behaviors must be detected in an early stage to prevent the system failure and SLA violation. In fact, the abnormalities are not necessarily anomalous but they can be an indication of new interesting behaviours, such as seasonal or new user behavior. These abnormalities refer to outliers. They are formally defined by Hawkis as [19]: *"An outlier is an observation which deviates so much from the other observations as to arouse suspicions that it was generated by a different mechanism"*. The process of discovering these outliers is referred to outlier detection. Practically, users emit a voluminous, unbound amount of requests coming continuously with a high rate and susceptible to certain changes over time simulating the behavior of the data streams. As a consequence, a perfect outlier detection should respect the data stream mining constraints, such as the real time mining, the limited time and memory requirements in addition to the adaptation to the concept drift. Therefore, it's essential to use outlier detection techniques in the context of data stream mining. Several works have been proposed in this context [8,9,18]. However, they are mainly built to detect the abnormal behaviors in the virtual machines. Contrary, we aim to predict the abnormality in the cloud servers by detecting the unusual tasks behaviors before their execution in the virtual machines. The contribution of this paper consists of the formulation of the users tasks and the server state as a data stream. Additionally, we attempt to create a framework that first permits the detection of the abnormality in the cloud users request in terms of CPU and memory by the use of AnyOut and MCOD outlier detection algorithms implemented in MOA framework. Second, it permits the comparison between them in this context in terms of the processing time, memory consumption and the effectiveness of the outlier detection. To accomplish that, we conducted several experiments applied on the data generated from CSG^+ which represents a cloud stream generator tool that we extended from CSG. Lastly, we have to mention that this work represents an extension of our precedent work [3].

This paper is structured into the following sections: Sect. 2 depicts the background of our article. Section 3 introduces the related works and the chosen approaches. Section 4 presents the proposed framework. Section 5 describes the conducted experiments. Finally, Sect. 6 concludes the paper with a summary and future work.

2   Background

2.1   Problem Presentation

The core of cloud computing consists of providing the users over the Internet with on-demand services. Mainly, there is three types of services which are the Infrastructure-as-a-service (IaaS), Platform-as-a-service (PaaS), and Software-as-a-serviceSaaS. These services represent the data center hardware and software. Actually, a cloud data center is equipped with a multitude of racks of servers hosting virtual machines that executes the user's request denoted by tasks and share the server capacity, which is materialized in term of resources (CPU, RAM, Bandwidth, etc.). These tasks represent the basic elements that consumes the resources since they are either storage or computation requests. The resource requirement of each task is materialized by the CPU, memory, bandwidth, network, etc. [1]. Generally, the bandwidth and network are uniform and do not have a big influence on the performance contrary to the CPU and memory. This problem will be formally described in the Sect. 2.2.

According to these facts, we conclude that the performance of the whole data center, especially the servers varies according to the user's requests and its available capacity. Thus, abnormal users requests behavior may threaten the whole data center and influence the services it provides. In this context, a perfect solution consists of using outlier detection techniques in the context of the data stream mining in order to detect these abnormalities in real time as the tasks arrive to prevent the system failure by predicting the state of the servers. The outlier detection in the context of the data stream mining is an interesting research area. It is emerging as a hot topic in many fields and application domains including intrusion detection, fraud detection, etc. It encompasses a broad set of techniques discussed in the Sect. 3.2.

2.2   Problem Statement

This problem statement is inspired from the work elaborated in [1,2]. The cloud is based on a set of data centers $DC = \{dc_1, ..., dc_m\}$ separated geographically and linked by the network. A data center dc_m is composed by a set of servers $S = \{s_1, ..., s_n\}$ that models the physical resources (Computing and storage).

$$dc_m \in DC, dc_m = \langle S \rangle \tag{1}$$

Each Server s_n contains a set of virtual machines $VM = \{vm_1, ..., vm_n\}$ and possesses certain resources.

$$s_n \in S, s_n = \langle VM_s, R_s \rangle \tag{2}$$

Each $vm_p \in VM$ executes a set of tasks $T = \{t_1, ..., t_i\}$ and requires some of the resources of server to perform.

$$vm_n = \langle s_n, T, R_v \rangle \tag{3}$$

A task $t_i \in T$ indicates the user requests and models the different cloud services. It's executed in the virtual machines and consumes resources materialized by CPU, memory, bandwidth, network, etc.

$$t_i = \langle vm_n, R_t \rangle \tag{4}$$

The Resource requirement of every component can be captured as vector R with p is a set of resources types; $p = \{CPU, memorey, bandwidth, I/O, etc.\}$.

$$R_s, R_v, R_t \subset R; R = \langle r_1......r_p \rangle \tag{5}$$

The users emit their requests(tasks) continuously with a high rate. Therefore, they are considered as a stream of data. It's crucial to detect the abnormal behavior in real time as data arrives.

At instant t the server receives a stream of tasks DS represented by their resource requirements where $d = p$. The resources are categorized according to the users requirements and the type of service they want.

$$DS = \langle ds_j^1, ds_j^2, ...ds_j^d \rangle$$

We aim to detect the outliers at instant t to predict the state of the server at instance $t + 1$. The general definition of an outlier in this context is considered to be the ds_j^d that shows different behavior from the rest of the dataset.

3   Related Work

3.1   Outlier Detection in the Cloud Computing

The outlier detection in the context of the cloud infrastructure is treated by various approaches. [18] presents an automated fault detection framework for cloud system. It uses the distance based technique destined for static data to identify the faulty machines basing on the list of machine performances provided by Ganglia monitoring system. [8,9] are based on the data stream mining techniques. In [8], a Chi-square test based anomaly detection framework using Apache Spark is proposed. It segments the stream of performance data (CPU and memory) of VMware virtual machines into windows of variable lengths and performs Chi-square based comparisons for anomaly detection. [9] is designed for cloud infrastructure providers to detect, during run time, unknown anomalies that may still be observed in complex modern systems hosted on VMs. This is achieved using an online clustering techniques where the normal behavior data are assigned to clusters located in a close neighborhood contrary to the anomalous ones. [22] predicts the performance anomalies in the virtual machines, in real time, using two-dependent Markov models used conjointly with Tree-Augmented Bayesian Networks. However, it requires labeled historical training data to derive the anomaly classifier. Thus, it can only predict the anomalies that the model has already seen before. The aforementioned approaches detect the abnormal behaviors within the virtual machines, during run time, in terms of the resource usage. While our approach intend to predict the servers behavior by detecting the abnormal tasks in terms of their resource demands, as they arrive, before their deployment in the virtual machines. Thus, this is can be considered to be more efficient since it predicts the occurrence of the abnormality in the server behavior in an early stage.

3.2 Outlier Detection in the Context of the Data Stream Mining

The outlier detection in the context of the data stream mining has recently reached popularity due to its effectiveness as well as the inherent importance of the data streams. For this reason, several research efforts have been conducted in the last decade until date. Many approaches have been proposed by adapting the traditional the outlier detection techniques to the context and the constraints of the data stream. Generally, the outlier detection techniques can be classified into different categories depending on the outliers definition. Examples of this include the statistical based technique that uses a data distribution model to declare points as outliers or inliers. The outliers have the lowest probability generated from the global distribution [8,11]. The distance based outlier detection technique decides the "outlierness" of an object based on the distances to its neighbors in the dataset [5,15]. This technique is not very efficient for regions with varying density, for that reason the density based technique was proposed. To detect outlier, this technique estimates the density around the data points. Contrary to normal points, outliers are surrounded by a density dissimilar to their local neighbor. The density is evaluated using an outlier score called LOF [13,14]. Clustering based outliers technique assumes that outliers either do not belong to any cluster, or situated far away from the center of their nearest cluster, or belong to small or sparse one. Different approaches based on clustering and its variant methods have been proposed [4,20]. The classification based technique classifies points either normal or outlier according to a model learned from training data. However, in case of the unsupervised classification an outlier boundary is learned around the normal data and points falling outside this boundary are considered as outliers [10,12]. The frequent pattern mining techniques construct a normal model by using the frequent patterns and assumes that the outliers are the patterns that do not conform to the established normal behavior [16,17]. Among these approaches, we chose two important ones; AnyOut and MCOD to compare their performances in our context.

AnyOut (Anytime Outlier Detection). [4] solves the anytime outlier detection problem for data streams with variable and constant arrival rate. It uses a cluster based approach to represent data in a hierarchical structure called clustree. An outlier score is calculated based on the deviation between the object and clusters in interruption by the next object. Furthermore, for constant data streams, AnyOut uses a FIFO queue or a sliding window to store objects and process them in parallel. Moreover, a confidence measure is used to control the distribution of the available time according to the complexity of the case.

MCOD (Micro-cluster-based Continuous Outlier Detection). [5] is a distance based algorithm that uses an event based approach. MCOD is based on the concept of evolving micro-clusters which correspond to regions containing inliers exclusively. The range queries for each new object are performed with respect to the micro-cluster centers in the current window w. An object p is an outlier if and only if there are less than k neighbors of a point p in all the inlier or potential outliers microclusters clusters, such that the distance between the centers of these micro-clusters is at most 3/2R.

4   Proposed Framework

In this section, we first justify the choice of the used outlier detection approches. Next, we descibe the proposed framework.

4.1   Choice of the Approaches

The choice of AnyOut and MCOD is due to different reasons: first, they both match with the problem characteristics in terms of different criteria (e.g. the computational cost, the nature of the input data (numeric). Second, AnyOut belongs to the clustering based technique which has a spread use [23]. In addition, it solves the problem of the anytime outlier detection. Thus, we want to evaluate its performance in our context. Third, MCOD represents a promising algorithm in the distance based approaches. It's characterized by a good efficiency and effectiveness [5]. An additional reason, is that both approaches are implemented within MOA framework. It gained a big popularity in the data stream mining. So, we are looking for exploring it and discovering its effectiveness.

4.2   Framework Description

The proposed framework permits the detection of the abnormal user's tasks coming to the cloud servers in terms of CPU and memory as defined in Sect. 2.2 using AnyOut and MCOD, in addition to the comparison between them in this context. It's composed of two models: the stream generation and the outlier detection. The former is elaborated using CSG+. The latter is established by the chosen algorithms implemented in the MOA framework. The output of this model is the performance associated to each algorithm such the processed time, the amount of memory required and the detected outliers. CSG^+ represents our extended version of CSG [1]. It is implemented in java and based on the "Cloudsim" simulator. It permits the creation of simulated data centers, servers, virtual machines and tasks. It outputs .arff files that represent the stream of tasks including their main features, such as resource requirement, length, etc. Its major importance resides in the ability of personalizing the generation process by the control of the speed, the concept drift, the number of tasks and virtual machines in addition to the their other characteristics. In order to efficiently observe the differences in the resource consumption, this generator gives the possibility to generate the tasks according to their resource consumption; small (small resource consumption), medium (medium resource consumption) and big (big resource consumption). The latter are further categorized into CPU intensive, memory intensive and Mix. This categorization together with the values of the resources are estimated from the google trace version 1 [6]. Google provided two trace versions. We chose the oldest one since it's simpler. It contains the normalized CPU and memory values of a set of tasks executing in Google within 7 h. The framework architecture is illustrated by the Fig. 1.

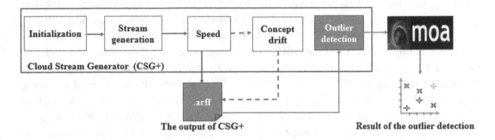

Fig. 1. Framework architecture

5   Experimentation

This section is devoted to present the experimentation applied to compare the chosen approaches. The first step to do is to introduce the experimentation set up. Next, we present the experimentation scenarios and results. At last, we finish by discussing the different results found.

5.1   Experimentation Setup

Data Set: In this experimentation, we generated two bunches of data from GSC^+. Each contains data sets of different sizes: $n = 50000, 100000, 150000, 200000$ and 250000. The first group consists of normal data free of outliers. The second one consists of data containing both normal and abnormal behaviors. Each instance in the data sets represents the task's CPU and memory. In practice, the tasks represent the servers workload. Generally, it's assumed that the small and medium tasks represent the majority of the data [21]. For that reason, we supposed that the big tasks are 30% of the whole data set.

Evaluation Metrics: Through this experimentation, we compared AnyOut and MCOD in terms of the time and memory efficiency and the outlier detection effectiveness. The effectiveness refers to the ability of the algorithm to detect the abnormal user tasks in terms of CPU and memory. It is evaluated using the false alarm rate which refers to the proportion of the normal tasks classified wrongly as outliers. On another side, the detection rate that depicts the proportion of the tasks that are correctly classified as outliers over the total number of the actual outliers.

5.2   Experimentation Scenarios

Our experimentation is conducted in two scenarios. The first one, consists of evaluating the performance of MCOD and AnyOut with normal data set free of outliers. The aim of this scenario is to test the ability of the chosen detection approaches to well handle normal data behavior in different situations. First, we varied the data size n and we fixed the window size (ws) to 10%n in order to reveal their ability to scale with the increase of the data size. Second, we assessed the impact of the window size ws by varying it for each data set in this range [1%n, 5%n, 10%n, 15%n, 20%n]. The purpose of the second scenario is to test the algorithms over data compromising both normal and abnormal behaviors. For that reason, we injected in every data set different percentage of outliers O randomly in the range of [0.1%n, 1%n, 3%n, 5%n]. This choice was based on the idea that, typically, the outliers represents only a small amount of data ranging between 0.01% and 5%. First, we assessed the impact of the data size by varying n and fixing ws to 10% and O to 1%. Second, we exposed the effect of the window size for the different outlier setting in the smallest and the largest data set D50 and D250 with n = 50000 and n = 250000 respectively. The last experiments consist of revealing the impact of the outlier rate O. To do that, we varied O in D50 and D250 in the different windows settings. The results of these experiments are discussed in Sect. 6. In both scenarios, we approximated the distance threshold $D = 1.5$ based on the values in the data set and $k = 50$ for MCOD. As for AnyOut, we tested different values for each parameter and we didn't find a significant change. For that reason, we kept the default values provided by MOA. On another side, we varied the *trainingSetsize* according to the data set size { 4000, 7000, 9000, 10000 } for { 50000, 100000, 150000, 200000 and 250000 } respectively. All the experiments carried out along this work were preformed on a HP pavilion G6. It has 8 GB RAM. In addition, it is equipped with a 2.10 GHz CPU, i3 processor on Windows 7.

6   Results and Discussion

The current comparison aims to expose the general differences between AnyOut and MCOD when varying data size, window size and outlier percentage. In terms of the processing time, overall, MCOD outperforms AnyOut in the different experiments, with anomalous and normal data sets. This may be explained because MCOD is based on the notion of micro-cluster where only inliers are stored. Adding and removing data points from micro-clusters are very efficient and not consuming. However, AnyOut is based on a hierarchical indexing structure of micro-clusters (Clustree). Inserting, deleting and updating data together with calculating the outlier score within the index structure can be reduced. Yet, it's more expensive than the use of the micro cluster in MCOD. Additionaly, AnyOut requires a training phase, thus additional time is consumed. In fact, with bigger data set size and larger window size more data is available. More data incurs bigger number of inliers. As a result, more data is stored in MCOD micro-clusters resulting in less distance calculation. In the other side, AnyOut does not differentiate between outliers and inliers while treating the data. It calculates and assigns the outlier scores for each point and insert it into the tree structure. This may be an advantage of AnyOut comparing to MCOD when the number of outlier increases. In fact, a bigger number of outliers incurs less inliers to store in MCOD. Consequently, this contributes to more distance calculation. While, in AnyOut the processing time generally decreases with the presence of outlier which represents and interesting finding. Even though, MCOD still outperforms remarkably AnyOut. Figure 2 depicts the processing time of Anyout and MCOD with normal and anomalous data sets where ws = 10% and O = 1% for the anomalous data. Addionnaly, the results revealed that Anyout processing time decreases in D50 with the increase of the outlier rate, except when ws = 20%n. This behavior is present in an opposite way in D250 where AnyOut processing time decreases with large windows: ws = 20%n and 15%n.

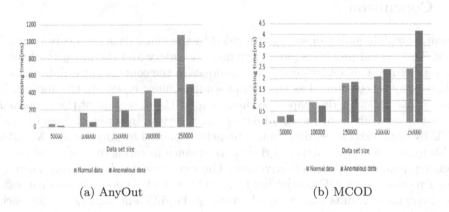

(a) AnyOut   (b) MCOD

Fig. 2. Processing time anomalous data sets vs Processing time normal data sets

In terms of the memory consumption, the increase of the data and window size in addition to the outlier rate yield to more consumption for both approaches. Additionally, AnyOut incurs better performance than MCOD in both scenarios except when n = 50000 and when ws = 20%n for the rest of the data sets. The consumed memory of MCOD corresponds to the memory required to store the information of each active object including the list of proceeding, the number of succeeding neighbors, the outliers of all the queries and the micro-cluster information. Therefore, bigger data and larger window size contributes to bigger amount of data to treat and more information to store. Moreover, larger outlier rate incures less data to store in the micro-cluster, more neighbors to store and more objects in the potential outlier list. All of this result in the increase of the memory consumption. As for Anyout. The memory consumed corresponds to the clusters features containing information about the objects together with a buffer to store the ones that cannot be inserted at a certain time.

The effectiveness decreases with the increase of the outlier rate and the size of the data and the window. Additionally, in the same way as the processing time and memory, MCOD gives the best performance. In contrast, AnyOut showed a very bad effectiveness manifested by a high false alarm rate for normal data and a low detection rate for the anomalous data. It showed a good effectiveness only with small data with small outlier rate $O = 0.1\%n$. Additionally we have to mention that both approaches did not rise any false alarm for the anomalous data set. MCOD has correctly classified on average more than 80% of the outliers for all the data sets. This represents a promising result. The effectiveness of MCOD may be explained by the continuous verification of the "outlierness" of the objects with each departure and arrival from the window. On the other side, the bad effectiveness of AnyOut might be justified by the interruption of the insertion every time an object arrives. Moreover, the level where the interruption occurs may affect the results if it doesn't give sufficient information.

7   Conclusion

Ensuring a high quality of services provided by the cloud data centers is a crucial task. Thus, cloud providers must guarantee the safety and the integrity of their infrastructure. The outlier detection techniques in the context of the data stream mining may be considered as an efficient solution since it detects the abnormalities in real time. In this paper, we investigated the detection of the abnormal users request coming to the cloud data centers servers in terms of memory and CPU by using two outlier detection approaches denoted by AnyOut and MCOD. Additionally, we have compared their performance in terms of processing time, memory requirement and effectiveness. The experiments showed very encouraging results for MCOD comparing to AnyOut that falls down in this context. However, these results can change by tuning the different parameters for each one of them. The choice of the parameters is a crucial condition that controls the results. This enhanced version of comparison will be done in further work. Another perspective of this work is the elaboration of a more excessive study of this problem in order to find the most suitable solution.

References

1. toumi, H., Brahmi, Z., Ben Arfa, Z.: Server load prediction using stream mining. In: The 31st International Conference on Information Networking (ICOIN) (2017)
2. Hassen, F.B., Brahmi, Z., Toumi, H.: VM placement algorithm based on recruitment process within ant colonies. In: The 1st International Conference on Digital Economy Emerging Technologies and Business Innovation (2016)

3. Souiden, I., Brahmi, Z., Toumi, H.: A survey on outlier detection in the context of stream mining: review of existing approaches and recommendations. In: The 16th International Conference on Intelligent Systems Design and Applications, Porto, Portugal (2016)
4. Assent, I., Kranen, P., Baldauf, C., Seidl, T.: AnyOut: anytime outlier detection on streaming data. In: The 17th International Conference on DASFAA, pp. 228–242 (2012)
5. Kontaki, M., Gounarisn, A., Papadopoulos, A.N., Tsichlas, K., Manolopoulos, Y.: Efficient and flexible algorithms for monitoring distance based outliers over data streams. Inf. Syst. **55**(3), 37–53 (2016)
6. Wilkes, J.: googleclusterdata. https://github.com/google/cluster-data (2013)
7. Zhang, Q., Cheng, L., Boutaba, R.: Cloud computing: state-of-the-art and research challenges. J. Internet Serv. Appl. **1**(1), 7–18 (2010)
8. Solaimani, M., Iftekhar, M., Khan, L.: Statistical technique for online anomaly detection using spark over heterogeneous data from multi-source vmware performance data. In: IEEE International Conference on Big Data (2014)
9. Sauvanaud, C., Silvestre, G., Kaaniche, M., Kanoun, K.: Data stream clustering for online anomaly detection in cloud applications. In: The 11th European Dependable Computing Conference, September 2015
10. Uddin, M.S., Kuh, A.: Online least-squares one-class support vector machine for outlier detection in power grid data. In: IEEE International Conference on Acoustics, Speech and Signal Processing (2016)
11. Uddin, M.S., Kuh, A., Weng, Y., Ilic, M.: Online bad data detection using kernel density estimation. In: IEEE Power and Energy Society and General Meeting (2015)
12. Kale, A., Ingle, M.D.: SVM based feature extraction for novel class detection from streaming data. Wirel. Personal Commun. J. **110**(9) 2015
13. Pokrajac, D., Lazarevic, A., Latecki, L.J.: Incremental local outlier detection for data streams. In: IEEE Symposium on CIDM, pp. 504–515 (2007)
14. Karimian, S.H., Kelarestaghi, M., Hashemi, S.: I-inclof: improved incremental local outlier detection for data streams. In: The 16th CSI International Symposium on Artificial Intelligence and Signal Processing (2012)
15. Cao, L., Yangt, Di., Wang, Q., Yu, Y., Wang, J., Rundensteiner, E.A.: Scalable distance based outlier detection over high-volume data streams. In: The 30th International Conference on Data Engineering (2014)
16. Lin, F., Le, W., Bo, J.: Research on maximal frequent pattern outlier factor for online high dimensional time-series outlier detection. J. Convergence Inf. Technol. **5**(10), 66–71 (2010)
17. Said, A.M., Dominic, P.D.D., Faye, L.: Data stream outlier detection approach based on frequent pattern mining technique. Int. J. Bus. Inf. Syst. **20**(1), 55–70 (2015)
18. Bhaduri, K., Das, K., Matthews, B.L.: Detecting Abnormal Machine Characteristics in Cloud Infrastructures. In: Proceedings of the IEEE 11th International Conference on Data Mining Workshops, pp. 137–144 (2011)
19. Hawkins, D.M.: Identification of Outliers. Springer, Heidelberg (1980)
20. Yogita, Toshniwal, D.: Unsupervised outlier detection in streaming data using weighted clustering. In: The 12th International Conference on ISDA, pp. 160–164 (2012)
21. Sheng, D., Derrick, K., Walfredo, C.: Characterization and comparison of cloud versus grid workloads. In: IEEE International Conference on Cluster Computing (2012)

22. Tan, Y., Nguyen, H., Shen, Z., Gu, X., Venkatramani, C., Rajan, D.: Prepare: predictive performance anomaly prevention for virtualized cloud systems. In: The IEEE 32nd International Conference on Distributed Computing Systems (ICDCS) (2012)
23. Vijayarani, S., Jothi, P.: Detecting outliers in Data streams using Clustering Algorithms. Int. J. Innov. Res. Comput. Commun. Eng. **1**(8) (2013)

Overlapping Community Detection Method for Social Networks

Mohamed Ismail Maiza[1(✉)], Chiheb-Eddine Ben N'Cir[2], and Nadia Essoussi[1]

[1] LARODEC Laboratory, ISG University of Tunis, Tunis, Tunisia
maizaismael@hotmail.fr, nadia.essoussi@isg.rnu.tn
[2] ESEN, University of Manouba, Manouba, Tunisia
chiheb.benncir@isg.rnu.tn

Abstract. A social network enables individuals to communicate with each other by posting information, comments, messages, images, etc. In most applications, a social network is modelled by a graph with vertices and edges. Vertices represent individuals and edges represent social interactions between the individuals. A social network is said to have community structure if the nodes of the network can be grouped into sets of nodes such that each set is densely connected internally. The investigation of the community structure in the social network is an important issue in many domains and disciplines such as marketing and bio-informatics. Community detection in social networks can be considered as a graph clustering problem where each community corresponds to a cluster in the graph. The goal of conventional community detection methods is to partition a graph such that every node belongs to exactly one cluster. However, in many social networks, nodes participate in multiple communities. Therefore, a node's communities can be interpreted as its social circles. Thus, it is likely that a node belongs to multiple communities. We propose in this paper a new overlapping community detection method which can be adopted for several real world social networks requiring non-disjoint community detection.

Keywords: Community detection · Overlapping clustering methods · Machine learning · Data mining · Social networks

1 Introduction

Social computing represents an important challenge for computing sciences. Social computing tasks have been the focus of many researchers in the last decades such as network modelling [7], centrality analysis [1], community detection [4], classification and recommendation [15]. Community detection is a crucial task of social computing which aims to find sets of nodes such that they are tied stronger to each other than to nodes outside each set. Due to the fast development of social networks, many lines of research have been introduced focusing on entities heterogeneity, the large size of social networks and the temporal development of social networks [22]. Community detection is applied in

© Springer International Publishing AG 2017
R. Jallouli et al. (Eds.): ICDEc 2017, LNBIP 290, pp. 143–151, 2017.
DOI: 10.1007/978-3-319-62737-3_12

many real world applications and it can be useful for other social computing tasks. Community detection methods deal with the issue of identifying communities in a network. Existing methods can be categorized into two types, disjoint and non-disjoint community detection methods. In the literature, disjoint community detection methods are classified into four main categories [19] which are node centric methods, group centric methods, network centric methods and hierarchy centric methods. Node centric methods require each node in a group to satisfy certain properties while community detection methods based on group-centric criterion consider the connections in the group in a way that it satisfies certain properties. Network-centric community detection methods focus on the global topology of a network. It aims to partition nodes of a into a number of disjoint sets optimizing a defined criterion over a network partition instead of one group. Hierarchical structure of communities are built based on network topology. There are two types of hierarchical clustering divisive, and agglomerative. There are also overlapping community detection methods, which aim to identify overlapping communities rather than disjoint communities. Several surveys were conducted on overlapping community detection methods. Clique percolation methods [22] are based on the assumption that a community consists of overlapping sets of fully connected sub-graphs and detects communities by searching for adjacent sub-graphs. Link partitioning methods [22] partition links instead of nodes in order to find overlapping communities. Local expansion and optimization methods aim to grow a community using a local benefit function. Fuzzy detection methods [22] quantify the strength of association between all pairs of nodes and communities using soft membership vectors. Agent based methods [22] identify overlapping community using processes in the graph. The described methods are either disjoint methods or non-disjoint methods. Disjoint methods are used when a disjoint partitioning is expected while non-disjoint methods are used when an overlapping cover is expected depending on the network. However, when there is expected shape of the partitioning, the expert can not determine what type of methods to use. We propose in this work to propose a new method that can detect disjoint communities and non disjoint communities based on a network centric method.

The remainder of this paper is structured as follows. In Sect. 2, the related works to traditional overlapping community detection and the background required about the network centric methods are described. In Sect. 3, the proposed method is detailed. An evaluation study to explore the performance of the proposed method will be explained in Sect. 4. Finally, in Sect. 5, some concluding remarks are given.

2   Related Works to Network Centric Community Detection

Network centric community detection methods deal with the problem of disjoint community detection from a point of view of the global topology of the network. Its goal is to group nodes of a network into a number of disjoint sets. These methods optimize a criterion defined over the network partition instead of over one group. Network centric community detection methods are modularity maximization, latent space models, block model approximation and vertex similarity. Modularity maximization [17] measures the quality of a community partition from a perspective of strength by considering the degree distribution of nodes. A network is characterized by a high modularity if there are dense connections between the nodes inside groups and sparse connections of nodes from different groups. Latent space models [12,13] aim to map nodes on a low dimensional euclidean space, thereafter the nodes are clustered using K-means clustering method [18]. The proximity between nodes defined on network connectivity are kept in the low dimensional space. Block model approximation [19] approximate a network by a block structure. The main principle is to identify entries that show an edge between two nodes in the adjacency matrix, and then approximate the adjacency matrix by a block structure where each block represents a community. Spectral clustering [20] came from the problem of graph partition. The goal of graph partition is to identify a partition where the number of links between two sets of nodes is minimized. Vertex similarity considers that the similarity of two nodes (vertex) is based on the similarity of their social circles defined by The number of connections they share in common [14]. There are many concepts that define vertex similarity, one of them is structural equivalence, where nodes V_i and V_j are structurally equivalent if they are connected to the same nodes. In other words, for any node V_k (different of V_i and V_j) that $V_k \neq V_i$ and V_j, $e(V_i, V_k) \in E$ iff $e(V_j, V_k) \in E$ (E is the set of edges and $e(V_i, V_k)$ is the edge between V_i and V_k). V_i and V_k are said to be equivalent if they are connected to exactly the same set of nodes.

When relations between nodes are represented as a matrix, columns of V_i and V_j are the same except for the diagonal entries. Nodes who share the same class of equivalence form a community. This measure requires nodes to be connected exactly to the same neighbours, which is practically uncommon. Alternatively, some simplified similarity measures can be used to measure the degree of equivalence between two nodes. Once a similarity measure is determined, classical k-means clustering method can be applied to find communities in a network [19]. This method detects only disjoint communities from a network.

3   Overlapping Community Detection Using Vertex Similarity

In this work, we aim to detect overlapping communities and disjoint communities depending on the applications requirements. To do so, we choose to assign a non disjoint partitioning method rather than the disjoint method as used to build clusters on the matrix of similarity. The main idea is to build a similarity matrix from the adjacency matrix, then apply Parameterized ROKM [5] which allows overlap size regulation, on this similarity matrix in order to detect overlapping communities based on the vertex similarity concept. Each case of the similarity matrix shows the degree of similarity between the node on the column and the node on the appropriate row. The similarity matrix is computed using Jaccard similarity [10]. Jaccard similarity measures the similarity of two nodes by the number of common neighbors they share. For two nodes V_i and V_j in the network, the Jaccard similarity between the two nodes is defined as:

$$Jaccard(v_i, v_j) = \frac{|N_i \cap N_j|}{|N_i \cup N_j|} = \frac{\sum_k A_{ik} A_{jk}}{|N_i| + |N_j| - \sum_k A_{ik} A'_{jk}} \tag{1}$$

where $|N_i|$ represents the cardinality of neighbors of the node V_i and k represents any node different from V_i and V_j. A_{ik} represents the connection between v_i and v_k (1 if an edge exists between nodes v_i and v_k and it is 0 otherwise). $|N_i \cap N_j|$ represents neighbours in common of V_i and V_j while $|N_i \cup N_j|$ represents the union of all the neighbours of V_i and V_j. The Jaccard similarity measure is within an interval [0.1]. The more neighbours V_i and V_j share, their Jaccard similarity increases. When the Jaccard similarity of two nodes is equal to one, the two nodes share exactly the same neighbours. Reciprocally, when their Jaccard similarity equals to zero, they have no common neighbours. For each node, we compute its Jaccard similarity with all the other nodes. As output we have a Jaccard similarity matrix where each node is represented by a vector of similarities with other nodes. After computing the similarity matrix, we use an overlapping clustering method on the Jaccard similarity matrix, Parameterized ROKM proposed by [5]. This method is an overlapping clustering method that allows overlap regulation based on the overlapping k-means [9] taking into account the number of assignment of the clusters. The main concept of the parameterized overlapping k-means is to assign a data observation V_i to the closest set of cluster profiles while minimizing its objective criterion. To decide whether to select a first cluster assignment combination A_i for the observationV_i or a second cluster assignment combination A'_i represented by binary vector. A positive difference in the produced local error is claimed. Parametrized ROKM controls the cluster assignment choice decision with the insertion of weights on the local errors relying on the size of the combinations. This weighting is manipulated using a parameter α. The assignment decision is formalized as follows:

$$\left\|A'_i\right\|^\alpha \cdot \left\|v_i - p^{A_i}\right\|^2 - \left\|A_i\right\|^\alpha \cdot \left\|v_i - p^{A'_i}\right\|^2 > 0 \tag{2}$$

where p^{A_i} represents the combination $p^{A_i} = \sum_k a_{i,k} p_{k,j} / \|A_i\|$ ($p_{k,j}$ is the profile of cluster j).

The objective criterion of $R_1 OKM$ based on the Overlapping K-means objective criterion:

$$J_{R_1 OKM}(A, P) = \sum_i \left[\sum_k a_{i,k} \right]^\alpha \sum_i \left(v_{i,j} - \frac{\sum_k a_{i,k} p_{k,j}}{\sum_k a_{i,k}} \right) \tag{3}$$

where k is the predetermined number of clusters. The model behaves in function of the value of α. If $\alpha = 0$, it extinguishes the effect of the weighting on the combination sizes. Therefore, when $\alpha = 0$, the obtained result is the same as the result of the OKM. When $\alpha > 0$, it restricts the assignment with wide combinations. As high as α rises, the overlaps reduce until the apparition of an identical clustering to the classic k-means. But when $\alpha < 0$, it supports the overlaps of the large assignments. As long as α bends down, the overlaps becomes large until the apparition of a clustering where all data objects are assigned to every cluster. The obtained cluster profile is a weighted mean of the cluster profiles according to each data belonging to the cluster. After applying the overlapping clustering method, each cluster represents a community.

4   Experiments

4.1   Datasets Description

In order to evaluate our method, we will use two real life datasets and a generated overlapping dataset of seven nodes. The real datasets are the Zachary's karate club [21] and the Football dataset [11]. Zachary's karate club dataset represents relations between 34 members of a karate club at a United States university where nodes represent members of the club, and edges (78) represents a friendship between two members of the club described by Fig. 2. The football dataset is a network of American football games between Division IA colleges during regular season Fall 2000 with 115 teams in 12 different conferences. Nodes represent teams while edges represent games between teams as shown in Fig. 1.

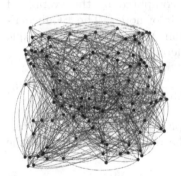

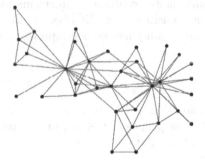

Fig. 1. The football dataset Zachary karate club network

Fig. 2. Zachary karate club network

4.2   Evaluation Measures

There are many evaluation measures which allow to compare the quality of clustering and community detection. In this work, since we perform an overlapping clustering method in order to detect overlapping communities, we use measures that compare between the result of an overlapping clustering method and a known ground truth dataset. Several metrics [6] were extended to be adapted for the evaluation of non-disjoint partitioning. Amigo et al. [2] proposed the BCubed metrics which aim to elaborate a refined comparison between overlapping partitioning. We detail in the following the extended BCubed metrics which we use to evaluate the performance of our proposed method. BCubed metrics [3] takes into account the multiplicity of shared communities between nodes. The computation of BCubed measures is based on the comparison of two partitions: a partition $C = \{c_1, ..., c_k\}$ defined by the ground truth which represents the true communities and a partition $P = \{p_1, ..., p_k\}$ of communities generated by the overlapping community detection method. The BCubed recall of an observation indicated how many nodes of its true community are assigned to the same community while BCubed precision shows how many nodes of its community belong to the same ground truth community.

$$Precision(N_i, N_j) = \frac{Min(|P(N_i) \cap P(N_j)|, |C(N_i) \cap C(N_j)|)}{|P(N_i) \cap P(N_j)|} \qquad (4)$$

$$Recall(N_i, N_j) = \frac{Min(|P(N_i) \cap P(N_j)|, |C(N_i) \cap C(N_j)|)}{|C(N_i) \cap C(N_j)|} \qquad (5)$$

where $P(N_i)$ and $C(N_i)$ are the sets of detected communities and ground truth communities respectively to which the observation N_i belongs to and $|. \cap .|$ the cardinality of shared communities. Multiplicity Precision is defined only if the observations N_i and N_j have some communities in common. Multiplicity recall is defined only if N_i and N_j have some ground truth communities in common. The global BCubed precision and recall can be computed by the average of all multiplicity recalls and precisions for every pair of nodes. The BCubed F-measure considers both BCubed recall and precision. We will use it as a unique measure to compare the overlapping community detection:

$$F_{measure} = \frac{1}{0.5 * \frac{1}{BCubed\,precision} + 0.5 * \frac{1}{BCubed\,recall}} \qquad (6)$$

In addition to BCubed measures, we also compare the size of obtained overlaps by the proposed method in this paper and the real overlap size one in the ground truth dataset:

$$Overlap\ size = \frac{1}{N} \sum_{i \in X} P(N_i) \qquad (7)$$

4.3   Results

The experiments were conducted with different values of α with k = 2 for the Zachary's and the generated datasets and with k = 12 For the football dataset. We compare our results with the Greedy Q method [16] and with ONDOCS [8] only for football dataset. As shown in Table 1, the proposed method is able to detect disjoint and non-disjoint communities depending on the value of the parameter α. For example, with Zachary's dataset, when $\alpha = 0.8$, the generated

Table 1. Comparison of the proposed method with on Greedy Q and ONDOCS

Dataset	Method	Size of overlap	F-measure
Generated	R-OKM ($\alpha = 0.5$)	1.14	0.864
	R-OKM ($\alpha = -0.5$)	1.14	0.864
Zachary's	R-OKM ($\alpha = 0.8$)	1	0.791
	R-OKM ($\alpha = -0.5$)	1.43	0.534
	Greedy Q	1	0.828
Football	R-OKM ($\alpha = 1.3$)	1	0.637
	R-OKM ($\alpha = 0.5$)	1.94	0.461
	ONDOCS	1	0.857
	Greedy Q	1	0.675

output is characterized with an overlap size of 1 which matches with the size of overlap of the ground truth set. When $\alpha = -0.5$ the generated overlap size is 1.43. With the football dataset, when α is high enough the overlap size is 1 (Disjoint communities) while with $\alpha = 0.5$ the overlap size becomes 1.94. This shows the ability of the proposed method to detect disjoint and non-disjoint communities from a network depending on the value of the parameter α. Despite the fact that our proposed method is able to detect disjoint and non-disjoint communities, unlike existing methods, it is clear from the results that the ONDOCS ans Greedy Q outperform the proposed method in term of quality. For example, with the football dataset, we obtained a value of F-measure of 0.637 while with Greedy Q 0.675 and with ONDOCS 0.857. This may be caused by the similarity measure which can be improved in future works.

5   Conclusion

In this paper we focused on overlapping community detection methods. Unlike traditional community detection methods, these methods detect disjoint communities without the possibility to allow to a node to belong to more than one community. We introduced a new community detection method that is able to detect disjoint and overlapping communities based on vertex similarity and a regulated overlapping clustering method. The performed experiments on a real and a generated dataset show the ability of the proposed method to detect overlapping and disjoint communities.

The proposed method can be extended in order to become able to detect the overlapping communities on large scale networks and from another perspective, the similarity measure can be improved in order to optimize the quality of the detected communities.

References

1. Agarwal, N., Liu, H., Tang, L., Yu, P.S.: Identifying the influential bloggers in a community. In: Proceedings of the 2008 International Conference on Web Search and Data Mining, pp. 207–218. ACM (2008)
2. Amigó, E., Gonzalo, J., Artiles, J., Verdejo, F.: A comparison of extrinsic clustering evaluation metrics based on formal constraints. Inf. Retr. **12**(4), 461–486 (2009)
3. Bagga, A., Baldwin, B.: Entity-based cross-document coreferencing using the vector space model. In: International Conference on Computational Linguistics, vol. 1, pp. 79–85 (1998)
4. Bedi, P., Sharma, C.: Community detection in social networks. Wiley Interdiscip. Rev. Data Min. Knowl. Discov. **6**(3), 115–135 (2016)
5. Ben N'cir, C.E., Cleuziou, G., Essoussi, N.: Generalization of c-means for identifying non-disjoint clusters with overlap regulation. Pattern Recognit. Lett. **45**, 92–98 (2014)
6. Ben N'cir, C.E., Cleuziou, G., Essoussi, N.: Overview of overlapping partitional clustering methods. In: Celebi, M.E. (ed.) Partitional Clustering Algorithms, pp. 245–275. Springer, Cham (2015). doi:10.1007/978-3-319-09259-1_8
7. Chakrabarti, D., Faloutsos, C.: Graph mining: laws, generators, and algorithms. ACM Comput. Surv. (CSUR) **38**(1), 2 (2006)
8. Chen, J., Zaïane, O., Goebel, R.: A visual data mining approach to find overlapping communities in networks. In: International Conference on Advances in Social Network Analysis and Mining, 2009. ASONAM 2009, pp. 338–343. IEEE (2009)
9. Cleuziou, G.: An extended version of the k-means method for overlapping clustering. In: 19th International Conference on Pattern Recognition, 2008. ICPR 2008, pp. 1–4. IEEE (2008)
10. Gibson, D., Kumar, R., Tomkins, A.: Discovering large dense subgraphs in massive graphs. In: Proceedings of the 31st International Conference on Very Large Data Bases, pp. 721–732. VLDB Endowment (2005)
11. Girvan, M., Newman, M.E.J.: Community structure in social and biological networks. Proc. Natl. Acad. Sci. **99**(12), 7821–7826 (2002)
12. Handcock, M.S., Raftery, A.E., Tantrum, J.M.: Model-based clustering for social networks. J. R. Stat. Soc. Ser. A (Stat. Soc.) **170**(2), 301–354 (2007)
13. Hoff, P.D., Raftery, A.E., Handcock, M.S.: Latent space approaches to social network analysis. J. Am. Stat. Assoc. **97**(460), 1090–1098 (2002)
14. Leicht, E.A., Holme, P., Newman, M.E.J.: Vertex similarity in networks. Phys. Rev. E **73**(2), 026120 (2006)
15. Liben-Nowell, D., Kleinberg, J.: The link-prediction problem for social networks. J. Assoc. Inf. Sci. Technol. **58**(7), 1019–1031 (2007)
16. Newman, M.E.J.: Fast algorithm for detecting community structure in networks. Phys. Rev. E **69**(6), 066133 (2004)
17. Newman, M.E.J.: Modularity and community structure in networks. Proc. Natl. Acad. Sci. **103**(23), 8577–8582 (2006)
18. Tan, P.-N., Steinbach, M., Kumar, V.: Introduction to Data Mining. Pearson Addison Wesley, Boston (2005). ISBN: 0-321-32136-7
19. Tang, L., Liu, H.: Community detection and mining in social media. Synth. Lect. Data Min. Knowl. Discov. **2**(1), 1–137 (2010)
20. Von Luxburg, U.: A tutorial on spectral clustering. Stat. Comput. **17**(4), 395–416 (2007)

21. Zachary, W.W.: An information flow model for conflict and fission in small groups. J. Anthropol. Res. **33**(4), 452–473 (1977)
22. Dourisboure, Y., Geraci, F., Pellegrini, M.: Extraction and classification of dense communities in the web. In: Proceedings of the 16th International Conference on World Wide Web, pp. 461–470. ACM (2007)

New Overlap Measure for the Validation of Non-disjoint Partitioning

Chiheb-Eddine Ben N'Cir[1,2](✉) and Nadia Essoussi[2]

[1] ESEN, University of Manouba, Manouba, Tunisia
[2] LARODEC, ISG, University of Tunis, Tunis, Tunisia
{chiheb.benncir,nadia.essoussi}@isg.rnu.tn

Abstract. Detecting overlapping groups is a specific challenge which offers relevant solutions for many application domains that require organizing data into non-disjoint clusters. Recently, several methods are proposed in the literature giving different layouts for the overlapping boundaries between clusters. However, the assessment process to evaluate the performance of these methods still a challenging issue to deal with. In fact, existing evaluation measures for overlapping clustering do not take into account the overlap error, local to each data object, while it calculates the whole overlap size relative to all clusters. Therefore, we propose in this work a new external evaluation measure, referred to as Micro-Overlap, able to perform efficient and robust evaluation of overlapping clustering. Experiments on synthetic and real datasets show the performance of the proposed measure compared to existing ones.

1 Introduction

Clustering is an important data mining technique which aims to group data by organizing similar data objects in shared groups based on their similar characteristics. Clustering has been applied in several domains such as customers segmentation [3], social networks [1] and image classification and segmentation. Usually, existing clustering methods in the literature lead to disjoint partitioning of data into non-connected groups. This fact may reduce quality of obtained clusters since several applications require organizing data into non-disjoint groups to fit the existing structures in data. An example of these applications is document clustering where a document may discuss several topics and then must be assigned to several topic categories (sport, politic, science, etc.). Another example of such specific application is emotion detection where a piece of music could evoke several emotions and therefore each song must be assigned to several emotions' categories (sad, quite, pleased, etc.). This kind of applications is referred to as Overlapping Clustering.

Overlapping clustering aims to organize data into non-disjoint groups by giving possibility for each data object to be assigned to many groups. This problem has been studied in the last few decades and several overlapping clustering methods are proposed which are based on different usual clustering approaches

© Springer International Publishing AG 2017
R. Jallouli et al. (Eds.): ICDEc 2017, LNBIP 290, pp. 152–161, 2017.
DOI: 10.1007/978-3-319-62737-3_13

such as hierarchical, graphical, generative, partitional, correlation and topological approaches. Nevertheless, few researches have studied the assessment of overlapping clustering which can be used to evaluate the quality of non-disjoint partitioning. Usually, the assessment process, also called validation process, is based on the comparison of an obtained partitioning and a known one given in prior (multi-labeled dataset). For example, a usual used evaluation is the comparison of the obtained size of overlaps by the clustering algorithm and the known one in the multi-labeled data set. More the size of obtained overlaps is near to the known one, more the clustering is considered better. However, this measure can lead to biased results since validation is considered for the whole dataset and does not take into account local error to each data object. To deal with this issue, we propose in this paper a new measure, referred to as "Micro-Overlap", for the assessment of overlap sizes in the context of overlapping clustering.

To summarize, our contribution to the literature consists in proposing a new *efficient* evaluation measure of overlap size which is able to give an accurate evaluation of *non-disjoint* partitioning. The remainder of this paper is structured as follows: Sect. 2 presents basic concepts of overlapping clustering methods. Then, Sect. 3 presents the process of quality assesments of overlapping clustering and describes the limits of using existing evaluation measures. After that, Sect. 4 describes the proposed Micro-Overlap measure to deal with the issue of quality assessment of overlapping clustering while Sect. 5 presents the different experiments which we have performed to show the effectiveness of the proposed measure. Finally, Sect. 6 gives concluding remarks and future works.

2   Overlapping Clustering

Overlapping clustering is based on the assumption that an observation may belong to one or several clusters. In this cluster configuration, obtained groups are usually non-disjoints and are called covers or clusters. A comparison of clusters obtained by hard and overlapping clustering is described in Fig. 1. This figure shows the non-disjoint partitioning where some data objects are assigned to more than one cluster. Several overlapping clustering methods based on hierarchical, graph, generative and partitional approaches are proposed in the literature.

Overlapping clustering methods, which are based on the partitional approach [10–12], are extensions of conventional partitioning methods such as k-means and k-medoids in which, overlaps between clusters are introduced in their optimized criteria. These methods can be divided into two groups: methods based on the additive model of overlaps and methods based on the geometrical model of overlaps. Examples of these methods are PCL [11], ALS [12], OKM [13] and R-OKM [10].

Besides the partitional approach, the graph approach was also used to build non-disjoint partitioning of data. Overlapping methods based on graph are mostly used in the context of community detection in complex networks [14,15]. In addition, the hierarchical approach is also used to build overlaps. Hierarchical overlapping methods propose a more richer visualization model to organize data.

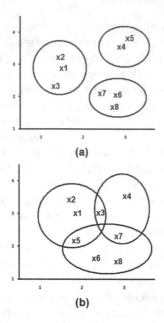

Fig. 1. Comparison of disjoint and non-disjoint clustering: (a) disjoint clustering and (b) non-disjoint clustering

Although the flexibility in visualization, hierarchical methods still too restrictive in overlaps while they do not study all the possible combinations of clusters for each observation. Example of hierarchical methods is the weak-hierarchies [17] method. Furthermore, the generative approach is also used to build overlaps. Generative overlapping methods [17,18] are extensions of the EM algorithm for non-disjoint partitioning of data. This category of methods assumes that each data is the result of a mixture of distributions.

All the described methods in this section deal with the issue of overlapping clustering and build non-disjoint partitionings of data. However, in real life application of overlapping clustering, user should choose the partitioning which fits the best the real structures that exist in data. So, how can we evaluate overlapping partitionings?

3   Motivation: Quality Assessment of Overlapping Clustering

The evaluation of clustering, also referred to as cluster validity, is a crucial process to assess the performance of the learning method in identifying relevant groups. This process allows the comparison of several clustering methods and allows the analysis of whether one method is superior to another one. Usually, the evaluation of clustering can be categorized into *internal* and *external* [6]. The first type, internal evaluation, is based only on the output of clustering by measuring the closeness of observations from one cluster and the distinctions of observations from different clusters. External evaluation, on the other hand, is based on comparisons between the output of the clustering and a dataset with known labels, also referred to as gold standard, usually built using human assessors.

For overlapping clustering, most of the validity measures traditionally used for clustering assessment, including both internal and external evaluations, become obsolete because of the multiple assignments of each observation. Despite this, some works propose an extension of well known validation measures to validate overlapping partitioning. In particular, internal evaluation measures, such as purity and entropy-based measures, cannot capture this aspect of the quality of a given clustering solution because they focus on the internal quality of the clusters. However, external validation measures, essentially Precision-Recall measures, were designed for overlapping partitioning in order to check the performance of the learning method in identifying non-disjoint clusters such as Pair based, BCubed [6] and Cice BCubed evaluations [9]. Another interesting external measure is the "Overlap size" which compares the obtained size of overlaps with a known size. More the value is close to the actual size of overlaps more the partitioning is considered better. The presented work focuses on issues related to the validation of overlapping clustering based on the overlap size measure.

In fact, the evaluation of overlap sizes is a necessary requirement to assess the quality of non-disjoint partitioning. This measure is used in several works [6,8,10] to evaluate the obtained sizes of overlaps compared to the known ones. Formally, the Overlap measure can be defined by:

$$Overlapsize = \frac{\sum_{i=1}^{k} |C_i|}{N} = \frac{\sum_{i=1}^{N} |A_i|}{N}, \tag{1}$$

where $|C_i|$ the number of data objects in the ith cluster, $|A_i|$ the cardinality assignments of each data object x_i, N the number of data objects and k the number of clusters. Overlap size is calculated by the average of assignments of all objects in the data set. For example, an overlap size equal to 2 indicates that, in average, each data object is assigned to two clusters. The overlap size is considered good when the value of overlaps of the obtained partitioning is near to the value of overlaps in the data set with known labels.

However, the assessment of non-disjoint partitioning based the overlap size measure may lead to biased evaluation of non-disjoint partitioning. Figure 1 presents a comparison between the overlap size of two different overlapping partitionings. Each partitioning represents the distribution of 8 data objects on the 3 clusters C_1, C_2 and C_3. This figures shows that the two partitionings are different: for example in the first partitioning (Fig. 2a), the third and the fourth data object are assigned to all the clusters while it is only assigned to the second cluster in the second partitioning (Fig. 2b). Although the existing differences between the two partitionings, obtained sizes of overlaps are equal evaluated by 1.5. Therefore, these two distributions have the same size of overlaps although

	Cluster 1	Cluster 2	Cluster 3
1	1	0	0
2	1	0	0
3	1	1	1
4	1	1	1
5	0	0	1
6	0	0	1
7	0	1	0
8	0	1	0

	Cluster 1	Cluster 2	Cluster 3
1	1	1	0
2	1	1	0
3	0	1	0
4	0	1	0
5	0	1	1
6	0	0	1
7	1	1	0
8	1	0	0

(a) Overlap size= 1.5   (b) Overlap size= 1.5

Fig. 2. Comparison of the overlap size of two overlapping partitionings: although the two partitionings are different, they have the same size of overlaps

they are totally different. This example shows the limit of using the overlap size measure which does not give an efficient evaluation of overlaps in the context of overlapping clustering. We show, in the next section, how can we design an efficient evaluation measure of overlap sizes.

4   Proposed Overlap Measure for an Efficient Evaluation of Overlapping Clusters

In order to deal with the evaluation of overlaps for overlapping clustering, we propose a local evaluation process, rather than the global existing one, which can evaluate overlaps for each data object separately. Then, local errors are averaged to build a global evaluation criterion. The proposed measure is referred to as "Micro Overlap" measure.

Given a known partitioning having the cardinality assignment $\{|B_i|\}_{i=1}^N$ of each data object x_i and a partitioning obtained by the clustering method having the cardinality assignment $\{|A_i|\}_{i=1}^N$ of each data object x_i, Micro Overlap begins by calculating the local overlap error of each data object x_i separately as described in Eq. 2.

$$Overlaperror_i = \frac{||A_i| - |B_i||}{max(|A_i|, |B_i|)} \quad (2)$$

where $|A_i|$ and $|B_i|$ are the cardinality of the set of clusters to which the object x_i is assigned to in the two compared partitionings. The local overlap error estimates the error of overlaps of each data object based on its cardinality in the two partitionings. Overlap error takes values between 0 and 1. A smaller value of this measure represents a better clustering: 0 indicates that object x_i has the same size of overlap in the two partitionings while 1 indicates the maximum of difference between the compared partitionings. In addition, the defined local error takes into account the difference that may exists in the number of clusters between the compared partitionings. It gives more importance to shared overlaps when the number of clusters is small rather than large shared overlaps when the number of clusters is large as described in the denominator of Eq. 2. This defined denominator maintains also the normalization of the overlap error (between 0 and 1) in order to easily interpret results.

Then, to build a global criterion for the evaluation of overlaps, we averaged overlap errors for all the data objects in the partitionings as follows:

$$MicroOverlap = \frac{\sum_{i=1}^{N} Overlaperror_i}{N} \qquad (3)$$

5   Experiments and Results

To evaluate the efficiency of the proposed overlap measure, we performed experiments on both simulated and real multi-labeled data sets.

5.1   Simulated Data Set

We build a test data set containing eight data objects organized into two non-disjoint clusters. We consider in the first partitioning (known labels) an overlap size equal to 1.125 where only the fourth data object is assigned to both clusters. Then, we build two other partitionings: in the first, only one data object is assigned to both clusters which is fifth data object and the second partitioning is the same partitioning of the known labels. Figure 3 shows the different described partitionings. We show in this figure that overlap size is the same for all the partitioning although partitioning 1 is different from the other partitionings. However, using micro overlap measure we show that micro-overlap is 0 only when partitionings are similar and is equal to 0.125 for the first partitioning. These obtained results show the effectiveness of using Micro-overlap rather than overlap size to evaluate overlapping clustering.

5.2   Real Multi-labeled Data Sets

We performed experiments on four real multi-labeled datasets having different degree of overlaps. Table 1 summarizes the statistics of the used datasets.

To build non-disjoint clusters with different degree of overlaps between clusters, we used the R1-OKM method [10] which provides a regulation of overlaps using a parameter $\alpha \in \mathbf{R}$. More the value of α is large, more overlaps are reduced until no-overlap between clusters. For each data set, we used three values of α:

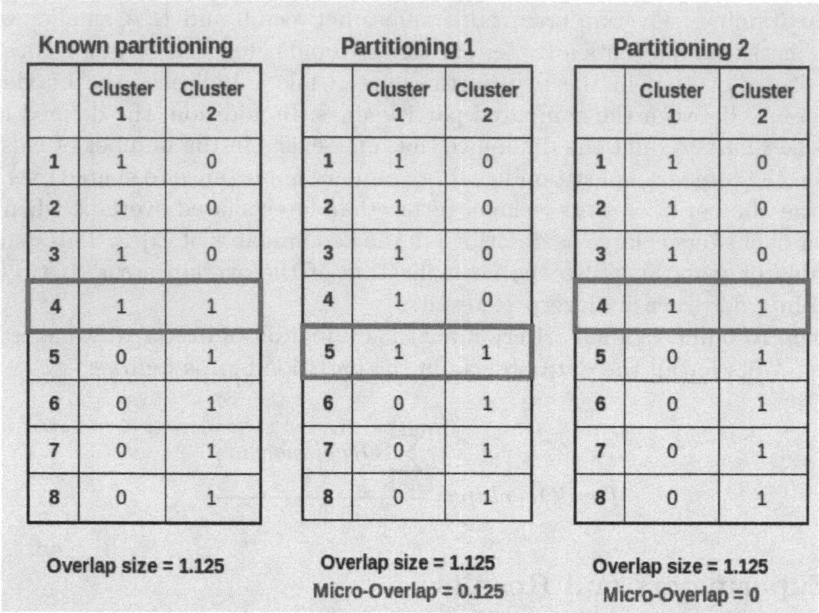

Known partitioning		
	Cluster 1	Cluster 2
1	1	0
2	1	0
3	1	0
4	1	1
5	0	1
6	0	1
7	0	1
8	0	1

Overlap size = 1.125

Partitioning 1		
	Cluster 1	Cluster 2
1	1	0
2	1	0
3	1	0
4	1	0
5	1	1
6	0	1
7	0	1
8	0	1

Overlap size = 1.125
Micro-Overlap = 0.125

Partitioning 2		
	Cluster 1	Cluster 2
1	1	0
2	1	0
3	1	0
4	1	1
5	0	1
6	0	1
7	0	1
8	0	1

Overlap size = 1.125
Micro-Overlap = 0

Fig. 3. Comparison of the overlaps using Overlap size and the proposed Micro-Overlap measures

Table 1. Data sets characteristic

Data set	Instances	Attributes	Classes	Overlap size
Emotion	593	72	6	1.81
Each movie	75	3	3	1.14
Yeast	2417	117	14	4.23
Scene	2407	300	6	1.07

−0.3, 0 and 2 and we considered a number of clusters equal to the number of classes in the labeled datasets. In order to compare the effectiveness of the proposed Micro-Overlap with the existing Overlap measure, we report BCubed precision, BCubed Recall and BCubed F-measure. These measures are usually used to compare the quality of overlapping clustering [6]. Obtained results are described in Table 2. These results firstly show the limit of using Overlap measure as a validation measure for overlapping clustering. For example in the Emotion dataset, which has an actual overlap evaluated by 1.8, obtained overlap size is evaluated by 2.11 which is relatively near to the actual overlaps. However, we show that obtained F-measure for this partitioning is poor (0.45) which do not indicates a good quality of partitioning. This problem is solved using Micro-Overlap measure which reports a high overlap error (0.31) for this partitioning. In fact, we can interpret from these results that overlapping data objects, in the

Table 2. Evaluation of partitionings obtained with R-OKM method using different values of α on real multi-labeled data sets

		Value of α		
		$\alpha = -0.3$	$\alpha = 0$	$\alpha = 2$
Emotion dataset	Precision	0.37	0.40	0.51
	Recall	0.58	0.53	0.2
	F-measure	0.45	0.46	0.28
Real overlap = 1.81	Overlap	2.11	1.93	1.05
	Micro.Overlap	0.31	0.30	0.36
Each movie dataset	Precision	0.19	0.26	0.56
	Recall	0.97	0.96	0.65
	F-measure	0.33	0.41	0.60
Real overlap = 1.14	Overlap	2.12	1.92	1.18
	Micro.Overlap	0.42	0.39	0.15
Scene dataset	Precision	0.006	0.12	0.46
	Recall	0.99	0.94	0.41
	F-measure	0.01	0.18	0.43
Real overlap = 1.07	Overlap	4.24	2.62	1.01
	Micro.Overlap	0.72	0.54	0.03
Yeast dataset	Precision	0.152	0.66	0.803
	Recall	0.87	0.36	0.01
	F-measure	0.25	0.46	0.02
Real overlap = 4.23	Overlap	8.4	4.1	1.00
	Micro.Overlap	0.52	0.33	0.72

real labeled data, are not the same considered as overlapping data objects in the obtained partitioning. The existing overlap measure cannot report this type of error.

Secondly, obtained results show that the proposed Micro-Overlap measure can evaluate efficiently the overlaps for *small* and *large* non-disjoint partitionings as reported for Scene and Yeast dataset which have an actual overlaps of 1.07 and 4.24 respectively. In Scene dataset, when obtained F-measure is relatively good and built overlaps are also small a good Micro-overlap is obtained (0.03). However, in Yeast dataset which is characterized by high actual overlaps, a poor value of Micro-overlap is reported (0.33) although obtained overlaps by clustering are high and coincides with the actual ones.

6   Conclusion and perspectives

We focused on this work on the external validation of overlapping clusters. We deal with the issue of the evaluation of overlap sizes when such non-disjoint partitioning is expected. We propose the Micro-Overlap measure which evaluates overlaps locally, observation by observation, rather than using a global overlap size as for the existing measures in the literature. The effectiveness of the proposed measure is observed on both synthetic and real multi-labeled datasets.

An interesting direction for future works is to integrate the evaluation of overlap sizes in the reported external validation measures such as precision, recall and F-measures. This could help users to easily interpret the quality of overlapping partitionings.

References

1. Wang, Q., Fleury, E.: Uncovering overlapping community structure. Complex Networks **16**, 176–186 (2011)
2. Zhang, H., Fritts, J.E., Goldman, S.A.: Image segmentation evaluation: a survey of unsupervised methods. Comput. Vis. Image Underst. **110**, 260–280 (2008)
3. Sarstedt, M., Mooi, E.: Cluster analysis. A Concise Guide to Market Research. STBE, pp. 273–324. Springer, Heidelberg (2014). doi:10.1007/978-3-642-53965-7_9
4. Steinbach, M., Karypis, G., Kumar, V.: A comparison of document clustering techniques. In: KDD Workshop on Text Mining, pp. 525–526 (2000)
5. Tsoumakas, G., Spyromitros-Xioufis, E., Vilcek, J., Vlahavas, I.: Mulan: a java library for multi-label learning. J. Mach. Learn. Res. **12**, 2411–2414 (2011)
6. Amig, E., Gonzalo, J., Artiles, J., Verdejo, F.: A comparison of extrinsic clustering evaluation metrics based on formal constraints. Inf. Retrieval **12**, 461–486 (2009)
7. Banerjee, A., Krumpelman, C., Ghosh, J., Basu, S., Mooney, R.J.: Model-based overlapping clustering. In: Proceedings of the Eleventh ACM SIGKDD International Conference on Knowledge Discovery in Data Mining, pp. 532–537 (2005)
8. Ben N'Cir, C.-E., Cleuziou, G., Essoussi, N.: Overview of overlapping partitional clustering methods. In: Celebi, M.E. (ed.) Partitional Clustering Algorithms, pp. 245–275. Springer, Cham (2015). doi:10.1007/978-3-319-09259-1_8
9. Rosales-Méndez, H., Ramírez-Cruz, Y.: CICE-BCubed: a new evaluation measure for overlapping clustering algorithms. In: Ruiz-Shulcloper, J., Sanniti di Baja, G. (eds.) CIARP 2013. LNCS, vol. 8258, pp. 157–164. Springer, Heidelberg (2013). doi:10.1007/978-3-642-41822-8_20
10. Ben N'Cir, C.E., Cleuziou, G., Essoussi, N.: Generalization of c-means for identifying non-disjoint clusters with overlap regulation. In: Pattern Recognition Letters, pp. 92–98 (2014)
11. Mirkin, B.G.: Method of principal cluster analysis. Autom. Remote Control **48**, 1379–1386 (1987)
12. Depril, D., Van Mechelen, I., Mirkin, B.G.: Algorithms for additive clustering of rectangular data tables. Comput. Stat. Data Anal. **52**(11), 4923–4938 (2008)
13. Guillaume, C.: Two variants of the OKM for overlapping clustering. Adv. Knowl. Disc. Manage. **2**, 149–166 (2009)
14. Gregory, S.: A fast algorithm to find overlapping communities in networks. In: Daelemans, W., Goethals, B., Morik, K. (eds.) ECML PKDD 2008. LNCS, vol. 5211, pp. 408–423. Springer, Heidelberg (2008). doi:10.1007/978-3-540-87479-9_45
15. Zhang, S., Wang, R.-S., Zhang, X.-S.: Identification of overlapping community structure in complex networks using fuzzy c-means clustering. Phys. A: Stat. Mech. Appl. **374**, 483–490 (2007)

16. Bertrand, P., Janowitz, P.F.: The k-weak hierarchical representations: an extension of the indexed closed weak hierarchies. Discrete Appl. Math. **127**, 199–220 (2003)
17. Heller, K., Ghahramani, Z.: A nonparametric Bayesian approach to modeling overlapping clusters. J. Mach. Learn. Res. **20**, 187–194 (2007)
18. Qiang, F., Banerjee, A.: Multiplicative mixture models for overlapping clustering. In: Proceedings of the IEEE International Conference on Data Mining, ICDM 2008, Washington, USA, pp. 791–796 (2008)

Uniformly Spread Embedding
Based Steganography

Marwa Saidi[✉], Houcemeddine Hermassi, Rhouma Rhouma,
and Safya Belghith

Laboratoire RISC, Ecole Nationale d'Ingénieurs de Tunis,
Université de Tunis El Manar, BP. 37, Le Belvédère, 1002 Tunis, Tunisia
marwoua.saidi@gmail.com

Abstract. In this paper, we propose a higher capacity steganographic
approach basing on a novel mapping method. We exploit the strong
energy compaction property in the Discrete Cosine Transform (DCT) to
ensure a maximum embedding capacity resulting minimum embedding
artifact. The approach combines the compaction efficiency with a ran-
domized embedding behavior generated by a chaotic function PieceWise
Linear Chaotic Map. The scrambled embedding positions ensures the
security aspect. The mapping applied over bits enlarges the capacity of
our approach by embedding symbols within DCT coefficients instead of
bits.

Keywords: High payload steganography · Frequency domain · Discrete
Cosine Transform · Scrambled embedding · Chaotic mapping

1 Introduction

Securing and protecting exchanged information between two communicators still
one of the major interest of researchers. Steganography has been introduced as
the science of hiding secret communication between a sender and a recipient by
ensuring the innocence aspect of this communication as well as its undetectabil-
ity. In fact, this science is based on three main axis: Payload, Security and
Robustness. The efficiency of the hiding approach (texts, images, audio, videos,
Internet packets) depends on the amount of distortion caused by the embedding
phase. Several steganographic approaches have been presented regardless to the
domain where the exchanged information is processing. In literature, a variety
of spatial steganographic approaches have been presented such as the Least Significant Bit (LSB) substitution where the secret bits are randomly positioned
in the least significant bits of the original pixels within the cover image [1–3].
As well as for frequency domain where a set of transform based steganography
approaches have been discussed such as DWT based steganography [4,5] or DCT
based steganography [6,7] or a combination of both transform [8].

This work mainly researches a novel mapping steganography approach based
on the Discrete Cosine Transform. We exploit the energy compaction property

© Springer International Publishing AG 2017
R. Jallouli et al. (Eds.): ICDEc 2017, LNBIP 290, pp. 162–172, 2017.
DOI: 10.1007/978-3-319-62737-3_14

characterizing the previously mentioned transform. Such decomposition allows us to embed the communicated secret message in the less detectable spots, in other words, coefficients that approximately has no impact over the quality of the original support. Since our approach requires an evaluation in terms of statistical properties, we use the ensemble classifier [14] and a set of features extractors that will be introduced later in steganalysis section.

The present paper is organized as follows. In Sect. 2, we provide a brief description of the used chaotic function. Then, Sect. 3 introduces the set of conventions used in the rest of the paper. The detailed steps of the proposed approach are presented in Sect. 4. All of the numerical results appear in Sect. 5 which includes a diverse set of steganalysis features in JPEG domain. Section 6 summarizes the present paper and outlooks the open challenges.

2   Randomized Embedding: PieceWise Linear Chaotic Map (PWLCM)

The $PWLCM$ [9] function is a nonlinear system generated by iterating Eq. 1. It has a chaotic behavior for an initial condition $x_0 \in [0,1]$ and a control parameter $\mu \in [0,1]$. This function is used as a generator of scrambled embedding positions in our proposed scheme.

$$x(n) = F[x(n-1)] = \begin{cases} \frac{x(n-1)}{\mu} & \text{if } 0 \leq \text{x(n-1)} \leq \mu \\ [x(n-1) - \mu]\frac{1}{0.5-\mu} & \text{if } \mu \leq \text{x(n-1)} \leq 0.5 \\ F[1 - x(n-1)] & \text{if } 0.5 \leq \text{x(n-1)} \leq 1 \end{cases} \qquad (1)$$

3   Conventions

We assumed that the sender's and the receiver's systems are configured with similar parameters. Initially a secret key denoted as K has to be exchanged. K contains $(x_0, \mu, \kappa, \eta, \delta)$. Throughout the following sections, we use the following notational conventions:

- Upper case letters denotes images in spatial presentation such as: $X = \{X_1, X_2,....,X_{M \times N}\}$ denotes an (M×N) sized grayscale cover image.
  $Y = \{Y_1, Y_2,....,Y_{M \times N}\}$ denotes an (M×N) sized grayscale stego image.
  $S = \{S_1, S_2,....,S_{m \times n}\}$ denotes an (m×n) sized grayscale secret image.
  $\widehat{S}$ denotes the extracted secret image.
- Lower case letters denotes images in transform domain such as x, y, s.
- $l \times l$ denotes the size of blocks.
- η: the number of AC coefficients holding the secret symbols in each block.
- δ: embedding factor used in mapping step.
- κ: the number of bits per symbol.

4   Proposed Method

Our approach is based on embedding secret symbols within DCT coefficients instead of embedding secret bits [10]. The embedding process starts with selecting an optimal value of the embedding factor δ which validates the extraction of the image in the reception. Then a mapping function is applied over the sequence of secret bits to be communicated: we divide the cover support into blocks, we apply the DCT transform, A generating process of chaotic positions within each block of coefficients is determined by $PWLCM$ function (Sect. 2), then the hiding process will take a place within scrambled spots where alteration is less detectable. The block diagram of our proposed data hiding algorithm is shown in Fig. 1.

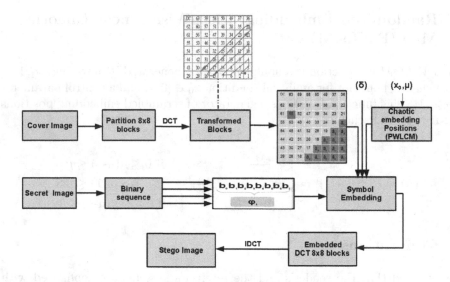

Fig. 1. Block diagram of the proposed embedding scheme.

4.1   Embedding Algorithm

- **Step1:** Divide X into a non overlapping blocks of $(l \times l)$ pixels $\{B_{X_i}\}_{i \in [1,L]}$, where L is the total number of divided blocks given by:

$$L = (\frac{M \times N}{l \times l}) \tag{2}$$

- **Step2**: Convert the secret image $S=\{S_1, S_2, ..., S_{m \times n}\}$ into its binary presentation denoted as B of length $|B| = 8 \times m \times n$.

$$B = b_1, b_1, ..., b_{|B|} \tag{3}$$

– **Step3:** we apply the following transformation composed of 4 consecutive sub-steps: first of all we transform the spatial blocks to frequency domain (Eq. i) then we effect to η ac coefficients an initial value (Eq. ii). Noting that the scan phase of ac coefficients in each block B_{x_i} starts from the least significant coefficient to the most significant one in a zigzag path. As a following stage, we apply the inverse transformation that we used previously (Eq. i^*) and we end up by converting again to the frequency domain where new η AC coefficients are estimated (Eq. ii^*).

$$
\begin{cases}
B_{x_i} = DCT_2(B_{X_i})_{i \in [1,L]} & (i) \\
B_{\tilde{x}_i} = Estimate(B_{x_i})_{i \in [1,L]} & (ii) \\
B_{\hat{X}_i} = IDCT_2(B_{\tilde{x}_i})_{i \in [1,L]} & (i^*) \\
B_{\hat{x}_i} = DCT_2(B_{\hat{X}_i})_{i \in [1,L]} & (ii^*)
\end{cases}
$$

– **Step4:** Apply a symbol mapping over the binary sequence, κ bits per symbol, by following Eq. 4. κ is variable in order to embed as maximum bits as we can. As a result we obtain a sequence of symbols $\{\varphi_i\}_{i \in [1,(|B|/\kappa)]}$.

$$
\varphi_i = (\alpha_i - (2^{(\kappa-1)})) * \delta \tag{4}
$$

with α_i denotes the decimal presentation of the κ bits.

Symmetric symbols upon zero are more flexible when it comes to extraction process, cause the receiver himself creates intervals of potential symbols using δ with regards of course to an error rate of extracted symbol detection. That's why we proceed the obtained vector of symbols to a supplement step.

$$
\{if \quad \varphi_i >= 0, \quad \varphi_i = \varphi_i + \delta; \tag{5}
$$

– **Step5:** Generate η chaotic embedding positions in each DCT block by iterating the $PWLCM$ function.
– **Step6:** One symbol is embedded per coefficient (Algorithm 1). The obtained steganographic blocks in frequency domain are denoted by $\{B_{y_i}\}_{i \in [1,L]}$. Finally we transform the $\{B_{y_i}\}_{i \in [1,L]}$ into the spatial domain. The resulting stego image to be transmitted is denoted by:

Algorithm 1. Symbol embedding

```
while k <= size(φ) do
    foreach block B_{x_i} do
        for j = 1 to η do
            B_{y_i}(j) = B_{x̂_i}(j) + φ_k;
```

$$
Y = \cup_{i \in [1,L]}(IDCT_2(B_{y_i})) \tag{6}
$$

4.2 Extraction Algorithm

The present process is based on comparing the extracted symbols and classify them according to a set of thresholds. Then a process of estimation take a place to retrieve the binary sequence previously embedded. Mainly, we process a copy of the received stego image so that we determine initially an estimated version of the cover support. Then, fundamentally, the embedded message is retrieved through a step of comparison and classification. Noticeably, the same sequence of chaotic positions used in the embedding is generated by the receiver thanks to the exchanged secret key K. The block diagram of our proposed data extracting algorithm is shown in Fig. 2.

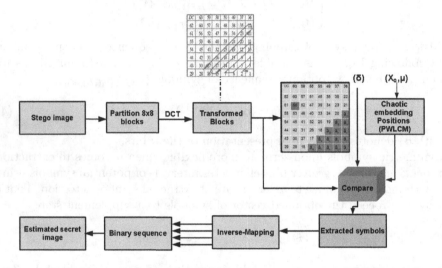

Fig. 2. Block diagram of the proposed extraction scheme.

- **Step1:** Divide the received stego image denoted as $\dot{Y}$ (Eq. 7) into a non overlapping blocks of $(l \times l)$ pixels.

$$\dot{Y} = \cup_{i \in [1,L]}(B_{\dot{Y}_i}) \tag{7}$$

- **Step2:** Estimate the cover support from the copy stego as a pretreatment phase by following the same steps described in Sect. 4.1 (step 3).

$$\begin{cases} B_{\dot{y}_i} = DCT_2(B_{\dot{Y}_i})_{i \in [1,L]} & (i) \\ B_{\tilde{\dot{y}}_i} = Estimate(B_{\dot{y}_i})_{i \in [1,L]} & (ii) \\ B_{\hat{Y}_i} = IDCT_2(B_{\tilde{\dot{y}}_i})_{i \in [1,L]} & (i^*) \\ B_{\hat{\dot{y}}_i} = DCT_2(B_{\tilde{\dot{Y}}_i})_{i \in [1,L]} & (ii^*) \end{cases}$$

- **Step3:** Generate the same η chaotic extraction positions in each DCT block by iterating the $PWLCM$ function.
- **Step4:** A simple comparison between η coefficients of the estimated cover blocks $\{B_{\hat{y}_i}\}_{i\in[1,L]}$ and the stego blocks $\{B_{y_i}\}_{i\in[1,L]}$ is effected. The difference between two coefficients denotes the embedded symbol. Later, the extracted symbol is sent as an input into a research function which provides as an output the corresponding binary presentation. The previously described function is based on the creation of a threshold table $T = [T_1, T_2, ..., T_{2^\kappa-1}]$ containing potential values of extracted symbols. The interval $[T_i, T_{i+1}]_{i\in[1,2^\kappa-1]}$ corresponds to one embedded symbol.
- **Step5:** Apply a κ bits inverse-mapping over each extracted symbol $\{\varphi_i\}_{i\in[1,(|B|/\kappa)]}$ using Eq. 8.

$$\alpha_i = \frac{\varphi_i}{\delta} + (2^{(\kappa-1)}) \tag{8}$$

With α_i is the decimal presentation of the corresponded κ embedded secret bits.
- **Step6:** Convert the extracted binary sequence $\widehat{B} = \{\widehat{b}_i\}_{i\in[0,|B|]}$ into an $m \times n$ image denoted by $\widehat{S}$.

5   Experimental Results

5.1   Framework

In this section we show the results of the proposed approach experimented using the following tools: Matlab R2015a and $AdobePhotoshopCS6$.

All experiments are conducted on images (covers and secrets) downloaded from the Standard test images database [11] and BossBase [12].

It contains 512×512 and 256×256, 8-*bit* grayscale uncompressed images, most of them are in TIFF (Tagged Image File Format) or PGM (Portable Gray Map) image format.

5.2   Comparison with Prior Works and Steganalysis

We define initially the embedding capacity (payload) of our proposed algorithm in Eq. 9 and we introduce the relative payload θ (Eq. 10) as the report between the number of messages that can be communicated (bits) and the size of the cover medium.

$$Payload = \log_2(|B|) \quad (bits) \tag{9}$$

$$\theta = \frac{\log_2 |B|}{M \times N} \quad (bpp) \tag{10}$$

With $|B| = |(0,1)|^{number of changes} = 2^{\kappa \times \eta \times L}$.

Since the importance of our approach is based on the effectiveness of using different mapping steps. We show in Fig. 3 the measured SSIM and PSNR over a

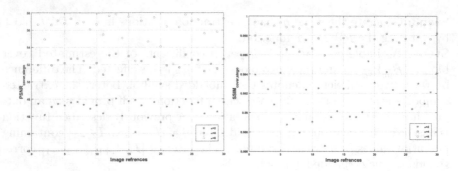

Fig. 3. Measured SSIM and PSNR for variable $\kappa = [2, 4, 8]$ and $\eta = [4, 2, 1]$.

set of 30 images. As we notice that both of the evaluation metrics are increasing proportionally to the mapping step. This makes perfect sense regarding to the fact that by increasing the number of mapped bits, we are altering less spots in the cover support which implies a better similarity between the original and the stego images.

Figure 4 presents the results of our approach for a set of cover objects: Peppers, Portofino, Babbon and Airplane of size 512×512 and secret image *Lena* of size 256×256. We measured the $PSNR$ between the cover and the stego objects as well as the PSNR between the original secret message embedded and the extracted one. We notice that the PSNR is acceptable (normally between 30 dB and 40 dB). For the following experiments we choose $\delta = 0.0788$, $\kappa = 8$ and $\eta = 16$. As it is mentioned in [13], applying steganography and steganalysis in real life scenarios is dealing with a lot of difficulties in terms of implementation. So, we present in this paragraph a statistical way of attacks based on a set of feature extracted from both cover and stego objects. Those features are feeded in a next step as an input into a binary classifier to verify the detectability of our steganographic approach in terms of minimum average probability of error P_E (Eq. 11).

$$P_E = \frac{1}{2} \times (P_{FA} + P_{MD}) \tag{11}$$

To ensure higher detection accuracy in digital steganography, we have to bear in mind the dependency of the feature extractor on the used steganographic approach. Feature-based steganalysis works systematically by adopting a specific model of the cover support then the steganalyzer's building process will take a place using machine learning algorithms.

In our case, we run our steganographic approach over a set of 30 cover and 30 stego images (15 Training/15 Testing) with $\theta = 0.125$ (*bpp*) taken from BossBase and converted into JPEG images with the quality factor 100. Since our approach is based on the alteration of the DCT Discrete Cosine Transform, we choose to evaluate the present embedding algorithm using 5 compatible feature extractors in JPEG domain with diverse dimensionality. Then to analyse the detection rate

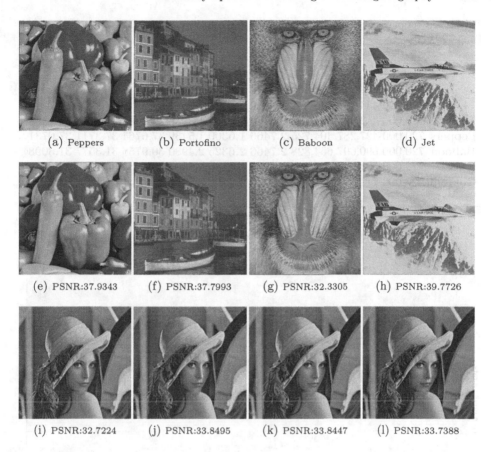

(a) Peppers   (b) Portofino   (c) Baboon   (d) Jet

(e) PSNR:37.9343   (f) PSNR:37.7993   (g) PSNR:32.3305   (h) PSNR:39.7726

(i) PSNR:32.7224   (j) PSNR:33.8495   (k) PSNR:33.8447   (l) PSNR:33.7388

Fig. 4. Embedding-Extracting results for $\delta = 0.0788$: Original Covers (a, b, c, d), Stego Images (e, f, g, h), Extracted Secret Images (i, j, k, l), $\eta = 16$, $\kappa = 8$, $l = 8$.

we feed the vector of features as an input into the ensemble Classifier with base learners implemented as Fisher Linear Discriminant (FLD) [14].

The set of feature extractors that we used to evaluate our approach contains mainly: Liu's 216-dimensional adaptive steganography-based features [15], CF^* [14] with a feature-space dimensionality = 7850, Holub et al.'s 8000-

Table 1. Measured P_E over 10 splits of the proposed approach for 5 features extractors.

Features	P_E									
LIU	0.0333	0.1000	0.0333	0.1000	0	0.1667	0.0667	0.0333	0.0667	0.1000
Chen	0.0333	0.0667	0	0.0333	0.0333	0	0	0.0667	0	0
CC-Chen	0	0	0	0.0333	0.0667	0	0	0.0667	0	0
DCTR	0.1000	0.0667	0	0.0333	0.0333	0.0333	0	0	0.0333	0.0333
CF^*	0.3000	0.2667	0.3000	0.3000	0.3333	0.3667	0.2333	0.4667	0.4333	0.3000

Table 2. Comparison of our approach ($\delta = 0.0788$, $\kappa = 8$, $\eta = 23$), Tang (2015) and Tang (2014).

Images	Payload(bits)			Payload(bpp)			PSNR		
	A1	A2	A3	A1	A2	A3	A1	A2	A3
Lena	720,000	434,130	410,871	2.7466	1.6561	1.5673	38.9301	37.8689	37.1282
Airplane	720,000	412,990	394,152	2.7466	1.5754	1.5036	39.8043	37.0148	36.7394
Peppers	720,000	427,287	404,725	2.7466	1.6300	1.5439	37.6764	37.3711	37.0731
Baboon	720,000	690,097	654,828	2.7466	2.6325	2.4980	30.9756	31.3577	31.3098
Boat	720,000	501,895	474,199	2.7466	1.9146	1.8089	38.9188	35.2349	34.9859
Goldhill	720,000	711,871	676,234	2.7466	2.7156	2.5796	37.8141	34.1212	33.7749

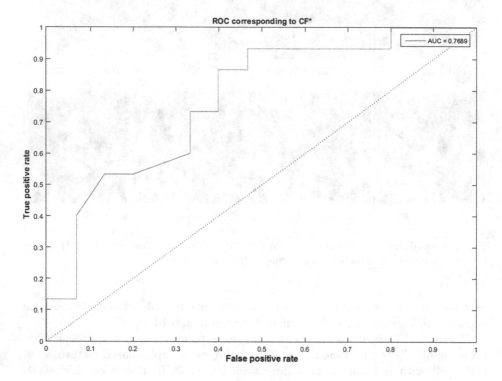

Fig. 5. ROC curve corresponding to CF^*.

dimensional DCT residuals-based feature (DCTR) [16], Chen [17] with feature-space of dimension = 486 and Chen enhanced by Cartesian Calibration's [18] 972-dimensional features. For each feature extracting method, we measured the P_E for the smallest payload $\theta = 0.125$, $\delta = 0.999$, $\kappa = 2$, $\eta = 4$ to check the robustness of our approach verses the ensemble analysis.

Table 1 shows the obtained P_E over 10 disjoint training and testing subsets (splits) for the 5 features extractors. Regarding to uniform embedding in real values of DCT coefficients, the probability of error $P_E \in [0, 0.1667]$ is dramatically low, for the first 4 features, which refers approximately to a perfect classification of our approach. However, it shows a better performance against ensemble analysis for the set of features provided using CF^*. We introduce in Fig. 5 the Receiver Operator Characteristics curve (ROC) corresponding to the Cartesian calibration feature set CF^*. We outline next in (Table 2) a comparison in terms of imperceptibility and embedding capacity of our approach $A1$ against two proposed schemes of Tang et al. ($A2$ [19], $A3$ [20]). Noticeably, our approach outperforms $A2$ and $A3$ with reference to the payload, however the PSNR measure remains approximately the same for all of the approaches.

6   Open Challenges

In this paper, we have proposed a new steganograghic approach based on DCT transform. We maximized the capacity of embedding by introducing the concept of mapped symbol instead of bits. The results have shown a better performance to previously proposed steganographic approaches.

A starting point in our future works can be the enhancement of the selection phase, where, the embedding artifacts are least detectable. In fact, we can present a distortion function applied over the cover. This distortion function mainly affect a block analysis over the DCT coefficient and determine the spots where a steganalyst has a minor probability to classify an object as stego.

Another two aspects that can be enhanced in our proposed work are, first, the idea of embedding the Dc and the most significant Ac coefficients of the secret image instead of embedding the whole pixels, and, second the possibility of shifting from the fixed embedding payload to the adaptive one. Another topic might be discussed in future propositions is "How much is it accurate to evaluate steganographic approaches that might be applied in real life scenarios using steganalysis tests with restriction to Lab's conditions?"

References

1. Yang, H., Sun, X., Sun, G.: A high-capacity image data hiding scheme using adaptive LSB substitution (2009)
2. Celik, M.U., Sharma, G., Tekalp, A.M., Saber, E.: Lossless generalized-LSB data embedding. IEEE Trans. Image Process. **14**, 253–266 (2006)
3. Wang, Z.-H., Chang, C.-C., Li, M.-C.: Optimizing least-significant-bit substitution using cat swarm optimization strategy. Inf. Sci. **192**, 98–108 (2012)
4. Baby, D., Thomas, J., Augustine, G., George, E., Michael, N.R.: A novel DWT based image securing method using steganography. Procedia Comput. Sci. **46**, 612–618 (2015)
5. Ghasemi, E., Shanbehzadeh, J., Fassihi, N.: High capacity image steganography using wavelet transform and genetic algorithm. In: Proceedings of the International Multiconference of Engineers and Computer Scientists (2011)

6. Li, X., Wang, J.: A steganographic method based upon JPEG and particle swarm optimization algorithm. Inf. Sci. **177**, 3099–3109 (2007)
7. Patel, H., Dave, P.: Steganography technique based on DCT coefficients. Int. J. Eng. Res. Appl. **2**, 713–717 (2012)
8. Kumar, V., Kumar, D.: Digital image steganography based on combination of DCT and DWT. In: Das, V.V., Vijaykumar, R. (eds.) ICT 2010. CCIS, vol. 101, pp. 596–601. Springer, Heidelberg (2010). doi:10.1007/978-3-642-15766-0_102
9. Mou, X., Li, S., Chen, G.: On the dynamical degradation of digital piecewise linear chaotic maps. Int. J. Bifurc. Chaos **15**, 3119–3151 (2005)
10. Saidi, M., Hermassi, H., Rhouma, R., Belghith, S.: A new adaptive image steganography scheme based on DCT and chaotic map. Multimedia Tools Appl. **76**, 1–18 (2016)
11. Standard test images database. http://www.imageprocessingplace.com/root_files_Vimage_databases.htm
12. Bas, P., Filler, T., Pevný, T.: "Break Our Steganographic System": the ins and outs of organizing BOSS. In: Filler, T., Pevný, T., Craver, S., Ker, A. (eds.) IH 2011. LNCS, vol. 6958, pp. 59–70. Springer, Heidelberg (2011). doi:10.1007/978-3-642-24178-9_5
13. Ker, A.D., Bas, P., Böhme, R., Cogranne, R., Craver, S., Filler, T., Fridrich, J., Pevný, T.: Moving steganography and steganalysis from the laboratory into the real world. In: Proceedings of the First ACM Workshop on Information Hiding and Multimedia Security, pp. 45–58 (2013)
14. Kodovsky, J., Fridrich, J., Holub, V.: Ensemble classifiers for steganalysis of digital media. IEEE Trans. Inf. Forensics Secur. **7**, 432–444 (2012)
15. Liu, Q.: Steganalysis of DCT-embedding based adaptive steganography and yass. In: Proceedings of the Thirteenth ACM Multimedia Workshop on Multimedia and Security, pp. 77–86 (2011)
16. Holub, V., Fridrich, J.: Low-complexity features for jpeg steganalysis using undecimated DCT. IEEE Trans. Inf. Forensics Secur. **10**, 219–228 (2015)
17. Chen, C., Shi, Y.Q.: JPEG image steganalysis utilizing both intrablock and interblock correlations. In: 2008 IEEE International Symposium on Circuits and Systems, pp. 3029–3032 (2008)
18. Kodovský, J., Fridrich, J.: Calibration revisited. In: Proceedings of the 11th ACM Workshop on Multimedia and Security, pp. 63–74 (2009)
19. Tang, M., Hu, J., Song, W., Zeng, S.: Reversible and adaptive image steganographic method. AEU-Int. J. Electron. Commun. **69**, 1745–1754 (2015)
20. Tang, M., Hu, J., Song, W.: A high capacity image steganography using multilayer embedding. Optik International Journal for Light and Electron Optics **125**, 3972–3976 (2014)

Uncertainty in Web Data

First Steps Towards an Electronic Meta-journal Platform Based on Crowdsourcing

Amna Abidi[1]([✉]), Nassim Bahri[1], Mohamed Anis Bach Tobji[1,2], Allel HadjAli[3],
and Boutheina Ben Yaghlane[4]

[1] ISG, LARODEC, Université de Tunis, Tunis, Tunisia
abidi.emna23@yahoo.fr, bahri.nassim@gmail.com
[2] ESEN, Univ. Manouba, Manouba, Tunisia
anis.bach@isg.rnu.tn
[3] ENSMA, LIAS, University of Poitiers, Poitiers, France
allel.hadjali@ensma.fr
[4] IHEC, LARODEC, University of Carthage, Tunis, Tunisia
boutheina.yaghlane@ihec.rnu.tn

Abstract. The last decade have witnessed a profusion of research work on the crowdsourcing topic. Human skills are essential in achieving high quality answers in crowdsourcing solving tasks. The current paper aims to introduce an innovative crowdsourcing-based solution for a scientific meta-journal. An overall architecture of the proposed system is introduced with a focus on the aggregation of the reviewers' evaluations to produce a final decision. We introduce several aggregation methods adapted to the nature of data to fusion and discuss them. In addition, we discuss future challenges that cope with the proposed system.

Keywords: Aggregation methods · Crowdsourcing · Possibility theory · Reliability · Human intelligent task · Information fusion

1 Introduction

Crowdsourcing consists in outsourcing tasks to distributed people and aims mainly at reducing cost and increasing quality of a final product or service. The last decade have witnessed a profusion of research works on the crowdsourcing topic [2,8,12]. Crowdsourcing key role is based on human skills in achieving high quality tasks. This concept, initially introduced as a business production model, is extended to other domains such as science.

In this paper, we introduce an electronic meta-journal platform (EMJ) based on crowdsourcing the reviewing task. In our proposed platform, workers are assumed to review submitted papers and to attribute a decision with an uncertainty measure [8]. We seek to investigate the aggregation of reviewers appreciations. Therefore, we present and discuss in this paper several aggregation methods where we consider workers' reliabilities, the decisions they produce and the possibility [13] they map to their decisions.

© Springer International Publishing AG 2017
R. Jallouli et al. (Eds.): ICDEc 2017, LNBIP 290, pp. 175–185, 2017.
DOI: 10.1007/978-3-319-62737-3_15

EMJ platform has several advantages. First of all, reviewing is done in triple-blind mode, where authors, reviewers and editors are all anonymous. This increases the objectivity of the final decision. Second, there is no assignment of reviewers. The selection of reviewers occurs "naturally" since they choose papers that fit with their domains. Third, as the platform is public, a great number of workers can handle one paper which gives an opportunity for authors to enhance their contributions. In addition, this fact makes the final decision more credible. Fourth, EMJ is a meta-journal, so the authors do not select a specific journal at the submission moment. If the article is accepted, it is assigned in a specific category. Reviewers help the system for that task. The out-of-scope problem is avoided[1]. Fifth, as the literature is plentiful of aggregation methods that enable to produce a final decision from reviewers' evaluations, the editor can choose the method he judges suitable, and can control, hence, the difficulty level of acceptance of its journal. Finally, the expertise of workers is not declared by reviewers themselves. The EMJ platform tries to estimate the reliability of workers according to objective measures such as bibliometrics.

However, many challenges have to be addressed; conflict of interest management, since the platform is public. Authenticity of reviewers and authors to avoid frauds[2]. Aggregation of reviewers evaluations. Estimation of workers' reliabilities. Workers motivation to incite scientists to review, etc. As stated above, we are focused in this paper in reviews aggregation.

The remainder of this paper is structured as follows: In the second section we present our background material; mainly the crowdsourcing concept, and the possibility theory. In the third section, we describe our system architecture. Section 4 provides the methods of aggregating reviewers appreciation. Section 5 is a discussion of the methods presented in Sect. 4. We show the remaining challenges of our system in Sect. 6. Finally, the paper is concluded in Sect. 7.

2   Background Material

2.1   Crowdsourcing

The Crowdsourcing can be defined as a business production model and distributed problem solving. Such model allows companies to reduce costs of solving some problems since tasks are distributed to networked people and not to the company's employees [12]. This outsourcing of tasks lets companies benefit from external expertise, creativity and collective intelligence of the crowd (network people) [9]. The term crowdsourcing was initially introduced in 2006 by Jeff Howe [6]. Since its emergence, many works were proposed and focused on different sides of crowdsourcing including applications, algorithms, quality management, cheating detection, etc. The advantages of crowdsourcing can be seen from a computational point of view where many trivial tasks for humans cannot

---

[1] In some journals, authors are informed that their submission is out of the scope several weeks after submission; a real waste of time for researchers.

[2] For example, an author that handles its own article with a second account.

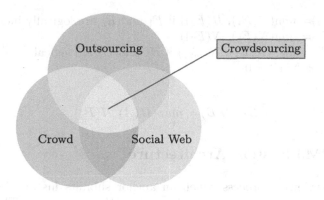

Fig. 1. The key elements of the Crowdsourcing (inspired from [10])

be resolved by machines. On the other hand, hiring employees to do this work will increase the costs of the company. One of the most popular and successful crowdsourcing platform is Amazon Mechanical Turk [1].

The crowdsourcing is the junction between the three elements depicted in Fig. 1 [10]: (i) Outsourcing (ii) Crowd (iii) Social Web. However, with the emergence of mobile, platforms of crowdsourcing are no more limited to web-based technologies and could be extended to run on mobile terminals [11].

Generally, crowdsourcing system contains 3 actors: the Requester (for whom the work is done), the crowd worker (who contributes in the work) and the platform used to manage the crowd or task.

2.2 Possibility Theory

The Possibility theory was introduced by [13] and developed by [3]. Probability concerns tendency of an event to occur while possibility concerns the ability of the event to occur. Possibility theory overcomes some drawbacks of the classical probability theory, mainly imprecision, partial ignorance and total ignorance.

A possibility distribution is a function $\pi: X \to [0, 1]$ and $\pi(a)$ expresses the degree to which a is a possible value for the considered variable. The normalization condition imposes that at least one of the values of the domain (a_0) is completely possible, i.e., $\pi(a_0) = 1$. The fundamental axiom is that the possibility of the disjunction of two propositions A and B is the maximum of the possibility of the individual propositions A and B:

$$\Pi(A \vee B) = max(\Pi(A), \Pi(B)) \tag{1}$$

Any event E is characterized by two measures: Its possibility Π (expressing the fact that E may more or less occur) and its necessity N (expressing that E will occur more or less for sure), $N(E) = 1 - \Pi(\overline{E})$.

- $\Pi(E_1 \cup E_2) = max(\Pi(E_1), \Pi(E_2))$

- $\Pi(E_1 \cap E_2)= \min(\Pi(E_1),\ \Pi(E_2))$ if E_1 and E_2 are logically independent.
- $N(E_1 \cap E_2)= \min(N(E_1),\ N(E_2))$
- $N(E_1 \cup E_2)= \max(N(E_1),\ N(E_2))$ if E_1 and E_2 are logically independent.
- $\Pi(E) < 1 \Rightarrow N(E) = 0$

$$\Pi(A \vee B) = max(\Pi(A), \Pi(B)) \tag{2}$$

3   The EMJ System Architecture

In classical reviewing process, when an author submits his article to a journal, the editor assigns a list of reviewers for the evaluation. The review is either blind where the reviewers know the authors' identities, or double blind where authors are anonymous. In both cases, authors have no idea about the reviewers' identities. Once the in-depth reviewing is finished, the editor summarizes the reviews and decides if the article should be accepted or rejected. This process supposes that editor is (1) available and can handle all papers at time, without blocking authors, (2) has a large network of reliable scientists who are available to review papers, (3) enough honest to assign objective reviewers without interfering in their decisions according to the authors identities (affiliation/race/country/religion etc.) and (4) able to aggregate the final decision from the reviewers' recommendations using a clear announced scientific method.

To cope with the aforementioned issues, we aim in this paper to introduce an innovative crowdsourcing-based solution to automate the reviewing task assignment and reviews aggregation. This solution reproduces the traditional reviewing process, but with two major modifications. The first one is that reviewers' assignment is no more the responsibility of the editor. Registered reviewers choose themselves papers for evaluation. The second change concerns the aggregation of reviews. At this stage, the received reviews are fused to get the decision about the acceptance/rejection of the paper following a scientific method.

The architecture of the proposed system is depicted in Fig. 2. It involves three actors: the author, the reviewers (crowd workers) and the editor. As we can observe, intervention of the editor is minor during the whole process. His main role is to configure the meta-journal. The system comprises three main sub-processes (numbered in Fig. 2) as follows:

1. *Submission:* It is the first step in our process. An author submits his article for evaluation. Therefore, an automatic check-style/template is performed and if there is no problem the article will be available for crowd workers for reviewing.
2. *Reviewing:* Reviewers choose the article if they think it fits with their research areas for evaluation. The system is configured to check the conflict of interest between the reviewer and the authors. Reviewers submit their evaluations, with a decision and a trust degree of their decision.

3. *Aggregating the reviews:* This task is automatically executed. It consists in collecting the reviews, and combining them to decide whether the article should be accepted or rejected. The decision is based on both the reviewers reliability and the trust degrees of their reviews (detailed in Subsects. 3.1 and 3.2).

The overall process is controlled by three important modules that impact the security (Authority control), the credibility (conflict management) and the reliability (Reliability estimation) of the final decision. These modules are detailed in Sect. 6.

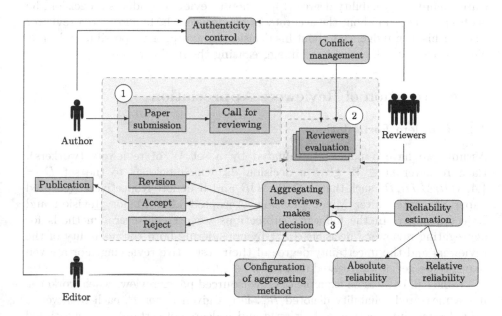

Fig. 2. EMJ system architecture.

3.1 Trust of the Reviewers

Crowd workers may have different levels of knowledge and expertise about the task to solve. Therefore, the reliability of the crowd workers is an important factor that needs to be addressed in crowdsourcing systems. Indeed, the quality of the result of a crowdsourcing task highly depends on the performance of the crowd workers [7,8]. For example, assume we have a paper entitled *Parallel K-prototypes for Clustering Big Data* that needs to be reviewed by the crowd workers of our system. A reviewer who has four papers about clustering of big data in serious conferences/journals should be more reliable than the one who has only one publication about the same topic.

In our approach, we consider the trust of a reviewer R_w as a combination between a *Relative Trust* (RT_w) and an *Absolute Trust* (AT_w). The absolute trust is derived from his bibliometrics (such as h-index, g-index etc.), his affiliation (ranking of his university), his works' history in the system (reviewers could receive feedbacks from authors for their reviews' quality), etc. The relative trust depends on the article/authors to review. It is a measure that is derived from (1) the matching between article field and the expertise field of the reviewers, and (2) the interest conflict degree between him and the paper authors.

3.2   Certainty of Reviews

In order to improve the pertinence of decisions made by the reviewers, we associate them to a possibility degree [13]. When a reviewer produces a decision, he is often not certain about the accuracy of his decision. EMJ allows to a reviewer to enter his own judgement about his decision by expressing a possibility degree. This measure will be considered in aggregating the final decision.

4   Aggregation of Reviewers Appreciation

4.1   Problem Formulation

Assume we have a paper P evaluated by a set W of reviewers (workers). Each reviewer $w \in W$ gives a decision d_w which belongs to the set $D = \{A, MR, MJR, R\}$ such that A, MR, MJR and R denote respectively $\underline{A}$ccepted paper, $\underline{A}$ccepted after $\underline{M}$inor $\underline{R}$evision, $\underline{A}$ccepted after $\underline{MaJor}$ $\underline{R}$evision and $\underline{R}$ejected paper. In the following subsections we present several methods for aggregating reviewers' decisions $\{d_w\}$ by considering both the reliability of the reviewers and the uncertainty degree of their respective reviewing. For a given paper P we get an aggregated decision $d^* \in D$.

Table 1 illustrates an example of crowdsourced paper reviews, each worker w has a measured reliability denoted R_w^P. For a given paper P, each reviewer w provides two information: a decision d and a degree of certainty π_w^P associated to his decision. Attributes S_{t-norm}, A_{t-norm} and L_{t-norm} are discussed later.

Table 1. Example of paper reviewing

W	d_w	R_w^P	π_w^P	S_{t-norm}	A_{t-norm}	L_{t-norm}
w_1	A	0.9	0.6	0.6	0.54	0.5
w_2	MR	0.7	0.8	0.7	0.56	0.5
w_3	MR	0.8	1	0.8	0.8	0.8
w_4	A	1	0.9	0.9	0.9	0.9
w_5	R	0.1	0.2	0.1	0.02	0
w_6	A	0.7	0.5	0.5	0.35	0.2

4.2 Majority Voting (MV)

In such method the final decision is claimed by the majority of the reviewers. Indeed, from a set of decisions D, the frequent decision d^* is selected to be the final one:

$$d^* = \max_{d \in D} \sum_{w \in W} \mu(d_w = d) \tag{3}$$

where $\mu(.)$ is a function which returns 1 if $d_w = d$ and 0 otherwise. From Table 1, we can conclude that the selected decision is A since it is the most voted. MV is considered as a naive method since it treats the worker's reviewing reliability as equal and does not distinguish the various skill level of workers. MV also ignores the uncertainty degrees assigned to workers' evaluations.

4.3 Confident Only Voting (CO)

For a submitted paper P, each reviewer decision d_w is associated with an uncertainty degree expressing how confident is his decision. It is denoted by π_w^P. Only possible (with $\pi_w^P = 1$) reviews are selected to determine the final decision d^*. However, to cover more reviews, we can use a fuzzy interval instead of certain possibility value (equal to 1 [2]). We define λ as the tolerance margin of the paper reviewing value denoted π_w^P. Indeed, we consider a review as certain if it belongs to the interval $[1 - \lambda, 1]$.

$$d^* - \max_{d \in D} \sum_{w \in W} \mu(d_w - d \wedge \pi_w^P \in [1 - \lambda, 1]) \tag{4}$$

As an example, we assume that $\lambda = 0.2$. The selected reviews from Table 1 are: $W_2 = MR(0.8)$, $W_3 = MR(1.0)$ and $W_4 = A(0.9)$. Hence, the final decision will be MR.

4.4 Confidence Weighted Voting (CW)

In our meta journal, the reviewers do not have the same profile neither the same confidence on their decisions. Instead of eliminating the reviewers who are not certain of their decision, as shown in the latter method, we set a confidence value to each worker. Our insight here is to give higher reliability denoted R_w^P to the evaluation of reviewers who are more trustworthy. Our method is inspired from the work of [2], the trust measure R_w^P is calculated from different aspects as presented in the Sect. 3.1.

$$d^* = \max_{d \in D} \sum_{w \in W} R_w^P \times \mu(d_w = d \wedge \pi_w^P \in [1 - \lambda, 1]) \tag{5}$$

4.5   Score-Based Method

This approach is derived from the work of Koulougli et al. in [8] where a score value is assigned to each review. The score calculation is based on aggregating functions borrowed from the fuzzy sets theory and known as T-norm (Triangular norms) operators. These operators are used to calculate the membership value of union, intersection and complement of fuzzy sets [4]. In this subsection, to calculate the score, we focus on the most common T-norms as follows:

- Standard intersection: $\forall(x, y) \in [0, 1]^2$, $T(x, y) = \min(x, y)$.
- Algebraic product (also called probabilistic operators [4]:) $\forall(x, y) \in [0, 1]^2$, $T(x, y) = x \times y$.
- Lukasiewicz t-norm: $\forall(x, y) \in [0, 1]^2$, $T(x, y) = \max(0, x + y - 1)$.

The last three columns in Table 1 represent the calculated score using the T-norms operators discussed above where S_{t-norm}, A_{t-norm} and L_{t-norm} denote respectively the Standard intersection, the Algebraic product and the Lukasiewicz t-norm. As shown in Table 1, S_{t-norm}, A_{t-norm} and L_{t-norm} returns the response (A, w_4).

5   Discussion

Although the aforementioned reviews aggregation methods meet our objective, each one has its advantages and limitations. We start with the Majority Voting which is very basic and simple to understand and to implement. However, it is a naive method which depends only on the number of occurrence of a response (in our case it is the decision) and don't consider the expertise of the reviewers (reliability) neither the uncertainty of their reviews. Therefore, using this method in our system will not make any sense and it will produce very approximate results. Unlike the Majority Voting, the Confident Only Voting takes into account the uncertainty degree of reviews. Indeed, only certain reviews are retained to determine the best decision. In the previous section, we extended the "certain case" of this method to cover more reviews with a margin of tolerance; we consider a review as certain if his possibility measure is in the interval $[1 - \lambda, 1]$. Despite the advantage of this method, it doesn't consider the reliability of the reviewers which has an important impact on the scientific quality of the final output. For example, if a reviewer whose reliability degree is 0.1 provides a review for which he assigns an uncertainty degree equal to 0.9, it will have a strong impact on our final decision since his expertise level is neglected. In fact, it could be interesting for journals that do not distinguish between the reviewers' skills and consider them all as equal. An alternative solution is presented in the Confidence Weighted Voting method. The novelty brought by this method is considering the reviewers' reliabilities. The insight of this method is to give higher weights to the reviews of reviewers who are more reliable. As our system is configurable, we can adjust the minimum value of reviewers' reliability (denoted min_{R_w}) according to the journal specification. For example, if an editor sets min_{R_w} to 0.8, that means

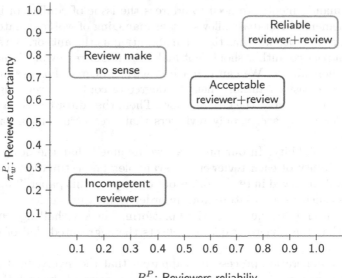

Fig. 3. Impact of π_w^p and R_w^p variations.

that only reviewers with a high level of expertise could evaluate the submitted papers. Figure 3 illustrates an example of possible scenarios depending on the variation of the two parameters λ and min_{R_w}. As we can see, the result quality depends from the values of λ and min_{R_w}.

Finally the last method is Score-Based which consists in assigning for each review a score depending on both the uncertainty of the response and the reliability of the reviewer. Therefore, the selected choice will be the one with the higher assigned score. Advantage of this method is considering all available information about the worker (reviewer reliability) and the work itself (review confidence), however, it presents two disadvantages: (i) There is only one person who decides and he is the one with the higher score, which means that this method doesn't support collaborative work (this is a need in our system) (ii) If we have two reviewers with a respective reliability value 0.8 and 0.2. The first reviewer gave to the possibility of the publication the value 0.2 and the second gave the value 0.9 then this paper will be accepted when we use the algebraic product or the Lukasiewicz t-norm.

6   Main Challenges

EMJ platform aims to automate crowdsourcing task of paper reviewing. The work presented in this paper is a first step toward a sophisticated crowdsourcing science platform. The main challenges for this system are:

- Authenticity of members: The membership process needs a robust proof of identity to avoid frauds.

- Conflict management: We need to address the issue of conflict of interest, a well-implemented Database allows better managing of workers matching. For example, a reviewer who has the same affiliation as the author, the same connection, a past co-author should not make the paper reviewing. This decision could be non binary. We can manage it with a degree. For example, if the reviewer is a past co-author, then the degree of conflict is very high. If he's from the same country, the degree is low. Then, the editor can set a threshold for this degree, to accept only reviewers that present a certain "amount of conflict".

- Reviewers reliability: In our proposal, we assumed that the degree of skills and uncertainty of each reviewer are available. The estimation functions of worker skills involved in team work is out of scope of this paper [2,5]. We need to address such issue in future work in order to measure reviewers reliability, involvement and task quality in EMJ platform. The key challenge comes from the fact that the reviewing quality reflects the aggregated skill of the entire team.

- Workers remuneration process: It is apparent that the involvement of reviewers in EMJ system contributes to the quality outcome of the task they undertake together. The crowd work in our platform needs to be remunerated in order to motivate reviewers to be involved in the process. Remunerating doesn't mean paying. Reviewers can obtain subscription in electronic journals, obtaining hard copies of volumes and issues, be funded for projects etc. A theoretical effort should be done in this way to correctly evaluate the work done by a reviewer.

- Paper reviewing could be open access to all reviewers member on EMJ platform. Some papers could have few evaluations. Others could be evaluated several times depending on the reviewers interest in such topic or availability. As a future work, we need to find a solution for delimiting the reviewing task of each paper.

7   Conclusions

In this paper, we have proposed an innovative crowdsourcing-based system for automating scientific paper reviewing. This electronic meta-journal platform consists of three components each one treats a specific sub-process. In this research project, we started by describing the architecture overview of our system. Therefore, we focused on the reviews aggregation which constitute the core of our platform. Thereafter, we presented several methods that fit with the problem in question.

The discussion showed that Confidence Weighted Voting may be the best method to aggregate the reviewers review since it considers both the reliability of the reviewers and the certainty of their reviews and highlights the cooperation of the reviewers team to produce a credible final result.

For future work, we plan to focus on estimation of the reviewers reliability, conflict management, reviewers remuneration and authenticity of the platform members.

References

1. Amazon mechanical turk. https://www.mturk.com/
2. Aydin, B.I., Yilmaz, Y.S., Li, Y., Li, Q., Gao, J., Demirbas, M.: Crowdsourcing for multiple-choice question answering. In: Proceedings of the Twenty-Eighth AAAI Conference on Artificial Intelligence, pp. 2946–2953. AAAI Press (2014)
3. Dubois, D., Prade, H.: Possibility Theory. Plenum Press, New York (1988)
4. Gupta, M.M., Qi, J.: Theory of t-norms and fuzzy inference methods. Fuzzy Sets Syst. 40(3), 431–450 (1991)
5. Rahman, H., Roy, S.B., Thirumuruganathan, S., Das, G., Amer-Yahia, S.: Worker skill estimation in team-based tasks, vol. 8, pp. 1142–1153, 11th edn. Association for Computing Machinery (2015)
6. Howe, J.: The rise of crowdsourcing. Wired Magaz. 14(6), 1–4 (2006)
7. Hung, N.Q.V., Thang, D.C., Weidlich, M., Aberer, K.: Minimizing efforts in validating crowd answers. In: International Conference on Management of Data, pp. 999–1014. ACM (2015)
8. Koulougli, D., Hadjali, A., Rassoul, I.: Leveraging human factors to enhance query answering in crowdsourcing systems. In: Tenth International Conference on Research Challenges in Information Science, pp. 1–6. IEEE (2016)
9. Pedersen, J., Kocsis, D., Tripathi, A., Tarrell, A., Weerakoon, A., Tahmasbi, N., Xiong, J., Deng, W., Oh, O., de Vreede, G.-J.: Conceptual foundations of crowdsourcing: a review of is research. In: 46th Hawaii International Conference on System Sciences, pp. 579–588. IEEE (2013)
10. Saxton, G.D., Oh, O., Kishore, R.: Rules of crowdsourcing: models, issues, and systems of control. Inf. Syst. Manage. 30(1), 2–20 (2013)
11. Yan, T., Kumar, V., Ganesan, D.: Crowdsearch: exploiting crowds for accurate real-time image search on mobile phones. In: 8th International Conference on Mobile Systems, Applications, and Services, pp. 77–90. ACM (2010)
12. Yuen, M.-C., King, I., Leung, K.-S.: A survey of crowdsourcing systems. In: IEEE Third Inernational Conference on Social Computing (SocialCom), pp. 766–773. IEEE (2011)
13. Zadeh, L.A.: Fuzzy sets as a basis for theory of possibility. Fuzzy Sets Syst. 1, 3–28 (1978)

Skyline Operator over *Tripadvisor* Reviews Within the Belief Functions Framework

Fatma Ezzahra Bousnina[1(✉)], Sayda Elmi[1,4], Mouna Chebbah[1,2],
Mohamed Anis Bach Tobji[1,3], Allel HadjAli[4], and Boutheina Ben Yaghlane[1,5]

[1] LARODEC, ISG, Université de Tunis, Tunis, Tunisia
fatmaezzahra.bousnina@gmail.com
[2] FSEGJ, Université de Jendouba, Jendouba, Tunisia
mouna.chebbah@fsjegj.rnu.tn
[3] ESEN, Univ. Manouba, Manouba, Tunisia
anis.bach@isg.rnu.tn
[4] LIAS, ENSMA, Université de Poitiers, Poitiers, France
{saida.elmi,allel.hadjali}@ensma.fr
[5] IHEC, Université de Carthage, Tunis, Tunisia
boutheina.yaghlane@ihec.rnu.tn

Abstract. The crowdsourcing *Tripadvisor* platform do not offer a multi-criteria filtering functionality for their users. Thus, these users are obliged to choose only one criteria to filter a query's results. In this paper, we introduce a new skyline operator, in the context of belief functions theory, to meet the multi-criteria filtering objective. The queried data, modeled with the theory of belief functions, takes into account all reviews and also reviewers' reliabilities. Experiments show interesting results of the proposed skyline operator in terms of size and performance.

Keywords: Evidence theory · Evidential databases · Reliability estimation · Combination · Discounting · Crowdsourcing · Skyline · Tripadvisor

1   Introduction

Crowdsourcing is a practice that asks the crowd or consumers via a questionnaire to propose or create a marketing policy. It provides a powerful system for creating data from real life participants. Partakers contribute with their feedbacks/reviews about a defined task. Crowdsourcing can be very challenging when it comes to gather and process information. *Tripadvisor* platform is one of the most well known crowdsourcing sites where travelers express their opinions about hotels they visited through an evaluation form. The collected reviews are used later to answer users' queries about the best hotels regarding some criteria like distance, price, etc. However, *Tripadvisor* can not answer to a multi-criteria query (for example, the cheapest and closest hotel to the beach). Skyline queries [2] are defined as preference queries that offer the possibility of multi-criteria filtering. Nevertheless, this kind of queries are not adapted to the crowdsourcing

© Springer International Publishing AG 2017
R. Jallouli et al. (Eds.): ICDEc 2017, LNBIP 290, pp. 186–197, 2017.
DOI: 10.1007/978-3-319-62737-3_16

platforms. They query databases where each tuple corresponds to a different hotel. In fact, they do not combine reviews about the same question. Thus, the use of the theory of belief functions to assess reviewers' reliabilities, to combine reviews and also to consider reviewers' reliabilities.

In this paper, we model the reviewers' feedbacks of *Tripadvisor* with the theory of belief functions [5]. First, we combine them to produce a data set of rates per hotel. Reviewers' scores are considered in the combination operation as the sources' reliabilities. Then, we introduce a new evidential skyline operator that deals with the particular type of obtained data. Finally, we implement the new operator and we lead experiments to compare its performance with the classic technique. Throughout this paper, example of Table 1 will be used. In this table, travelers give their evaluations about hotels $\{h_1; h_2; h_3; h_4\}$ over a scale of 6 notes.

In the sequel of this paper, some basic concepts of the theory of belief functions, evidential databases and skyline operator are presented in Sect. 2. Aggregation of travelers' reviews considering their reliabilities are presented in Sect. 3. In Sect. 4, the new skyline operator applied over the obtained *Tripadvisor* data is proposed. Experimental results and comparison with the classic evidential skyline [7] are also presented in the same section. Conclusion and future works are held in Sect. 5.

Table 1. Reviews about hotels

Reviewers	Hotels	Price	Place	Service	Score
R_1	h_1	3	4	3	1510
R_2	h_1	−1	4	2	22800
R_3	h_2	4	−1	5	400
R_4	h_2	3	5	−1	8140

2 Background Material

In this section, some basic concepts relative to the belief functions theory, evidential databases and skyline operator are presented.

2.1 Theory of Belief Functions

The theory of belief functions was introduced by Dempster and Shafer [5,6,11], it is also called *theory of evidence* or *the Dempster-Shafer theory*. In one hand, evidence theory provides an explicit representation of uncertainty and imprecision. In the other hand, it models other types of imperfection such the partial and the total ignorance. Let Θ be a finite set of exhaustive and mutually exclusive hypotheses called *frame of discernment*. The *power set* $2^\Theta = \{\{\varnothing\}, \{\theta_1\}, \{\theta_2\}, ..., \{\theta_n\}, \{\theta_1, \theta_2\}, .., \{\theta_1, \theta_2, ..., \theta_n\}\}$ includes all subsets of Θ. A *basic belief assignment* (*bba*), also called a *mass function*, is a mapping $m : 2^\Theta \longrightarrow [0, 1]$ such that

$$m(\emptyset) = 0 \quad \text{and} \quad \sum_{A \subseteq \Theta} m(A) = 1 \tag{1}$$

When $m(A) > 0$, A is called a *focal element*. The mass $m(A)$ is the belief committed exactly to A and to none of its subsets.

The *belief function*, denoted *bel*, represents the minimal degree of faith committed exactly to an hypothesis A, such that:

$$bel(A) = \sum_{B, A \subseteq \Theta : B \subseteq A} m^\Theta(B) \tag{2}$$

The *plausibility function*, denoted *pl*, is the maximal degree of faith committed to an hypothesis A, such that:

$$pl(A) = \sum_{B, A \subseteq \Theta : A \cap B \neq \emptyset} m^\Theta(B) \tag{3}$$

The belief function, *bel*, quantifies the degree of faith on a proposition A justified by degrees of supports (masses) of its subsets. It quantifies also the degree of faith on a comparison. Thus, comparing two independent probability distributions is easy in the framework of probability theory. However, standard *bel* and *pl* functions are not able to manage comparisons. Indeed, their definitions were extended [1,8,9] to meet the aim of comparing two independent (*bbas*).

Let X and Y be two independent variables $m_X, m_Y : 2^\Theta \rightarrow [0,1]$ their respective evidential values. The *bel* of inequalities are defined as follows:

Definition 1. *(Inequality bel(X ≤ Y))*

$$bel(X \leq Y) = \sum_{A \subseteq \Theta} (m_X(A) \sum_{B \subseteq \Theta, A \leq^\exists B} m_Y(B)). \tag{4}$$

Definition 2. *(Inequality bel(X < Y))*

$$bel(X < Y) = \sum_{A \subseteq \Theta} (m_X(A) \sum_{B \subseteq \Theta, A <^\forall B} m_Y(B)) \tag{5}$$

In the theory of belief functions, a large set of combination rules [10] merge *bbas* in the aim of improving decisions. The first one is the Dempster's rule of combination [5] that generalizes the Bayes rule. It is normalized and it combines mass functions produced from different and independent sources. The joint mass is obtained from merging two *bbas* using the orthogonal sum. This rule of combination is commutative, associative but not idempotent[1].

Definition 3. *Let m_1 and m_2 be two independent mass functions, the joint mass $m_{1 \oplus 2}$ is computed such that:*

$$m_{1 \oplus 2}(A) = \begin{cases} \dfrac{\sum_{B \cap C = A} m_1(B).m_2(C)}{1 - \sum_{B \cap C = \emptyset} m_1(B).m_2(C)} & \forall A \neq \emptyset \\ 0 & \forall A = \emptyset \end{cases} \tag{6}$$

---

[1] $m \copyright m \neq m$.

A particular combination is the discounting which considers sources' reliabilities into their mass functions. It is a specific mechanism to the belief functions theory that discounts masses proportionally to their sources' reliabilities. However, sources' reliabilities need to be learned before the discounting.

The reliability factor α in $[0, 1]$ characterizes the credibility of a source. Note that (i) $\alpha = 1$ represents a fully reliable source, (ii) $\alpha = 0$ represents an unreliable one and (iii) $1 - \alpha$ is the discounting. The discounted mass m^α is computed as follows:

$$\begin{cases} m^\alpha(A) = \alpha.m(A) & \forall A \subset \Theta \\ m^\alpha(\Theta) = \alpha.m(\Theta) + (1 - \alpha) \end{cases} \tag{7}$$

The theory of belief functions, is used to model imperfect data in many domains like medicine and weather forecasting. Such data need to be stored in order to be later queried. Thereby, specific database models that can handle data modeled with belief functions theory were introduced.

2.2 Evidential Databases

An *Evidential database* denoted (EDB), also named *Dempster-Shafer database*. The evidential database model was firstly introduced by Lee [8,9]. Later on, other models were proposed [1,3,4]. An EDB stores perfect and imperfect information, modeled using the evidence theory. It has N tuples and D attributes. An *evidential value*, denoted V_{ta} is the value of an attribute a for a tuple t that represents a *bba*, m_{ta}, such that:

$$V_{ta} : 2^{\Theta_a} \rightarrow [0, 1] \tag{8}$$

$$\text{with } m_{ta}(\varnothing) = 0 \text{ and } \sum_{A \subseteq \Theta_a} m_t(A) = 1 \tag{9}$$

The set of focal elements of a *bba* V_{ta} is noted F_{ta} such that:

$$F_{ta} = \{x \subseteq \Theta / m_{ta}(x) > 0\}$$

2.3 Skyline Operator

Skyline operator over an EDB introduced by [2] is based on the formal model of Pareto dominance also called Pareto preference.

Let $\mathcal{H}$ be a collection of objects defined on a set of attributes $A = \{a_1, a_2, \ldots, a_d\}$ such that:

Definition 4. *(Pareto Dominance). Given two objects $h_t, h_l \in \mathcal{H}$, h_t dominates h_l (in the sense of Pareto), denoted by $h_t \succ h_l$, if and only if h_t is as good or better[2] than h_l in all attributes and strictly better in at least one attribute, i.e., $\forall a_r \in A : h_t.a_r \leq h_h.a_r \wedge \exists a_\ell \in A : h_t.a_\ell < h_l.a_\ell$ where $h_t.a_r$ and $h_l.a_r$ stand for the rth attribute of h_t and h_l, respectively.*

---

[2] To make simple and without loss of generality, we assume through all the paper that the smaller the value the better it is.

Definition 5. *(Skyline). The skyline of $\mathcal{H}$, denoted by $Sky_{\mathcal{H}}$, includes objects of $\mathcal{H}$ that are not dominated by any other object, i.e., $Sky_{\mathcal{H}} = \{h_t \in \mathcal{H} \mid \not\exists\, h_l \in \mathcal{H}, h_l \succ h_t\}$.*

In this paper, we propose a new optimized evidential skyline operator that we apply over the *Tripadvisor* travelers' reviews; however, earlier these responses are treated with the belief functions' tools. In the next section, we present the modeling of given responses as *bbas* and then the discounting with sources' reliabilities. Finally, these discounted *bbas* are combined per attribute for the different hotels.

3   Elicitation of Reviewers' Feedbacks as Belief Functions

Tripadvisor provides a reviewing form for travelers in order to evaluate hotels according to several criteria. A response about one criteria for a specific hotel can be in $\{-1; 1; 2; 3; 4; 5\}$. A response in $\{1; 2; 3; 4; 5\}$ is precise and certain. It induces a precise and certain belief function. The response -1 reflects the total ignorance. All responses to the same review (same hotel and same criteria) are combined in order to provide one *bba* that summarizes all the reviewers' evaluations. Note that, responses need to be discounted to take into account reviewers' reliabilities. A reviewer response is translated into a *bba*, in the context of belief functions theory.

3.1   Construction of Mass Functions

Belief functions theory allows the construction of basic belief assignments (*bbas*) from the set of hypotheses. The mass of an hypothesis A as modeled in Eq. (1) and denoted, $m(A)$, is interpreted as the degree of support given by an expert and that reflects his belief on that response A. This mass can not be divided on subsets of A. In *Tripadvisor* platform, each traveler chooses one rate from 1 to 5. If he does not provide a rate, his response is interpreted as -1. From the theory of belief functions' point of view, the frame of discernment is $\Theta = \{1, 2, 3, 4, 5\}$. We recall that -1 is interpreted as total ignorance, $m(\Theta) = 1$. Each non empty response is interpreted as certain and precise belief functions over Θ.

Example 1. The first reviewer R_1 gives a rate 3 for the service of hotel h_1. This response is interpreted as $m(3) = 1$.

Table 2 is an interpretation of Table 1, in the context of belief functions theory for criteria: Price, Place and Service.

These mass functions are combined in order to have only one *bba* for each hotel. Before combining these reviews, they have to be discounted to take into account the travelers' reliabilities. Therefore, reviewers' reliabilities are firstly estimated.

Table 2. Construction of mass functions

Reviewers	Hotels	Price	Place	Service	Score
R_1	h_1	$m(3) = 1$	$m(4) = 1$	$m(3) = 1$	0.136
R_2	h_1	$m(\Theta) = 1$	$m(4) = 1$	$m(2) = 1$	0.99
R_3	h_2	$m(4) = 1$	$m(\Theta) = 1$	$m(5) = 1$	0.036
R_4	h_2	$m(3) = 1$	$m(5) = 1$	$m(\Theta) = 1$	0.733

3.2 Reliability Estimation and Discounting

One of the most interesting challenges in crowdsourcing is quantifying the reliability of reviewers. The conflict between two experts' opinions reflects the unre-
liability of at least one of them. The estimated reliability of each reviewer is
used later to weaken their given opinions modeled through the basic belief
assignments(*bbas*). The *Tripadvisor* platform attributes to each reviewer a number of points depending to its contributions. These points are accumulated when
the traveler (reviewer) gives an opinion about a hotel that he visited. Figure 1(a)
shows how the *Tripadvisor* rewards reviewers that add photos, videos, helpful
reviews, etc. Added to that, *Tripadvisor* divides its reviewers into 6 levels, shown
in Fig. 1(b): the first level is assigned to travelers having 300 to 2499 points and
the final and the sixth level is affected to travelers with points starting from
10.000. Method of rewarding travelers as illustrated in Fig. 1 is fixed by the
Tripadvisor platform.

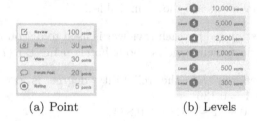

(a) Point   (b) Levels

Fig. 1. Computation of points in *Tripadvisor* and their corresponding levels

We propose to estimate the reliability of each reviewer based on points and
levels given by the *Tripadvisor* platform. Thus, we propose two methods: the
first is to calculate a reliability for each reviewer having points from 300 to 10.000
relatively to the sixth level, as shown in Eq. (10), and the second is to compute
the reliability score for reviewers having more than 10.000 point (i.e., travelers
that acquire the last level and accumulating more points), as shown in Eq. (11).

The maximal score is fixed to 0.9 for the 10.000 points. Based on that, a
reliability is computed for reviewers having points under 10.000, such that:

$$Score = (points * 0.9)/10.000 \qquad (10)$$

When the number of points accumulated by a reviewer are greater that 10000, the reliability is computed such that:

$$Score = 1 - (1/points) \tag{11}$$

Figure 2 shows the reviewers' reliabilities according to accumulated points.

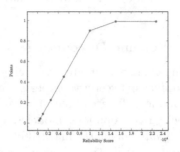

Fig. 2. Estimated reviewers' reliabilities

Example 2. the first reviewer R_1 in Table 1 has accumulated 1510 points and since his number of points is lower than 10000 then his reliability score is computed using method (i): $Score_{R_1} = 1510 * 0.9/10000 = 0.136$. The second reviewer R_2 has accumulated more points than 10000 then his reliability score is computed using method (ii): $Score_{R_2} = 1 - (1/22800) = 0.99$. Estimated reliabilities for all reviewers are shown in Table 2.

The reliability estimated for each reviewer is used to discount the basic belief assignments that reflect their reviews about hotels using Eq. (7).

Example 3. The reviewer R_1, the reliability degree is $\alpha = 0.136$. Thus:
$m^\alpha_{Price}(3) = 0.136 * 1 = 0.136$
$m^\alpha_{Price}(\Theta) = 0.136 * 0 + (1 - 0.136) = 0.864$

Results of discounted mass functions are shown in Table 3.

Once the reviews, modeled as *bbas*, are discounted, they may be combined.

3.3   Combination of Reviews

In theory of belief functions, combination rules aggregate data from different and independent sources to get one mass function that reflects all sources' opinions.

Example 4. Reviews about hotel h_2 for attribute *Price* are combined as shown in Table 4.

Table 3. Discounting of mass functions

Reviewers	Hotels	Price	Place	Service
R_1	h_1	m(3) = 0.136	m(4) = 0.136	m(3) = 0.136
		$m(\Theta) = 0.864$	$m(\Theta) = 0.864$	$m(\Theta) = 0.864$
R_2	h_1	$m(\Theta) = 1$	m(4) = 0.99	m(2) = 0.99
			$m(\Theta) = 0.01$	$m(\Theta) = 0.01$
R_3	h_2	m(4) = 0.036	$m(\Theta) = 1$	m(5) = 0.036
		$m(\Theta) = 0.964$		$m(\Theta) = 0.964$
R_4	h_2	m(3) = 0.733	m(5) = 0.733	$m(\Theta) = 1$
		$m(\Theta) = 0.267$	$m(\Theta) = 0.267$	

Table 4. Combination of *bbas* about the price of h_2

Price	$m_{R_3}^{h_2}(\Theta) = 0.964$	$m_{R_3}^{h_2}(4) = 0.036$
$m_{R_4}^{h_2}(\Theta) = 0.267$	$m_{34}(\Theta) = 0.26$	$m_{34}(4) = 0.01$
$m_{R_4}^{h_2}(3) = 0.733$	$m_{34}(3) = 0.7$	$m_{34}(\emptyset) = 0.03$

The joint mass of reviewers R_3 and R_4, $m_{3\oplus4}$ about the price of hotel h_2 is: (i) $m_{3\oplus4}(3) = 1/(1 - 0.03) * 0.7 = 0.72$; (ii) $m_{3\oplus4}(4) = 1/(1 - 0.03) * 0.03 = 0.012$; (iii) $m_{3\oplus4}(\Theta) = 1/(1 - 0.03) * 0.26 = 0.268$.

Similarly, we combine all *bbas* for each attribute for the different hotels. The obtained evidential database EDB is in Table 5.

The obtained database is evidential with either precise *bbas*, or partial ignorance *bbas*. This EDB is then queried with preference conditions using the skyline operator. Preference conditions may deal either with one attribute like Price, Place or Service or with a combination of these attributes leading to the multi criteria filtering.

Table 5. Evidential database

Hotels	Price	Place	Service
h_1	m(3) = 0.136	m(4) = 0.9814	m(2) = 0.98
	$m(\Theta) = 0.864$	$m(\Theta) = 0.0086$	m(3) = 0.01
			$m(\Theta) = 0.01$
h_2	m(3) = 0.72	m(4) = 0.992	m(5) = 0.036
	m(4) = 0.012	$m(\Theta) = 0.008$	$m(\Theta) = 0.964$
	$m(\Theta) = 0.268$		

4   Skyline Operator in Tripadvisor and Experimental Results

Applying the evidential skyline operator [7], we can apply an existing method according to the *Tripadvisor* data. The dominance relationship extended to evidential data can be defined as follows:

Definition 6 *(The b-dominance). Given two objects $h_i, h_j \in \mathcal{H}$ and a belief threshold b, h_i b-dominates h_j denoted by $h_i \succ_b h_j$ if and only if h_i is believably as good or better than h_j in all attributes a_r in A $(1 \leq r \leq d)$ and strictly better in at least one attribute a_{r_0} $(1 \leq r_0 \leq d)$ according to a belief threshold b, i.e., $\forall a_r \in A : bel(h_i.a_r \geq h_j.a_r) \geq b$ and $\exists a_l \in A : bel(h_i.a_l > h_j.a_l) \geq b$.*

Given an object h_i, we denote by $h_i.a_r^-$ and by $h_i.a_r^+$ respectively the minimum value and the maximum value of the *bba* defined on the attribute a_r denoted by $h_i.a_r$.

Property 1. Let b be a belief threshold, if the mass function affected to the partial ignorance of a *bba* $h_i.a_r$ is greater than (1-b), i.e., $m_{h_i.a_r}(\Theta) > (1 - b)$ then $bel(h_i.a_r \geq h_j.a_r) < b$. In this case, the object h_i can not b-dominate h_j.

Example 5. Suppose we have $b = 0.6$. Let $h_i.a_r$ and $h_j.a_r$ be two *bbas* defined on objects h_i and h_j, respectively, and defined on the attribute a_r such that $h_i.a_r = \{3\}(0.4), \{\Theta\}(0.6)$ and $h_j.a_r = \{2\}(0.3), \{\Theta\}(0.7)$. $bel(h_i.a_r \geq h_j.a_r) = 0.12 < 0.6$ since $m_{h_i.a_r}(\Theta) = 0.6 > (1 - b)$.

Property 2. Let b be a belief threshold, if $m(h_i.a_r^+)$ is greater than b, i.e., $m(h_i.a_r^+) \geq b$ then $bel(h_i.a_r \geq h_j.a_r) \geq b$.

Property 3. Let b be a belief threshold, if $m(h_i.a_r^-)$ is greater than b, i.e., $m(h_i.a_r^-) \geq b$ then $bel(h_i.a_r \geq h_j.a_r) < b$.

Intuitively, an object is in the believable skyline if it is not believably dominated by any other object.

Based on the b-dominance relationship, the notion of b-$sky_{\mathcal{H}}$ is defined as follows.

Definition 7 *(The b-skyline). The b-skyline of $\mathcal{H}$ denoted by b-$sky_{\mathcal{H}}$, comprises those objects in $\mathcal{H}$ that are not b-dominated by any other object, i.e.,*

$$b - sky_{\mathcal{H}} = \{h_i \in \mathcal{H} \mid \nexists\ h_j \in \mathcal{H}, h_j \succ_b h_i\}$$

Property 4. Given two belief thresholds b and b', if $b < b'$ then the b-$sky_{\mathcal{H}}$ is a subset of the b'-$sky_{\mathcal{H}}$, i.e., $b < b' \Rightarrow b$-$sky_{\mathcal{H}} \subseteq b'$-$sky_{\mathcal{H}}$.

Proof. Assume that there exists an object h_i such that $h_i \in b$-$sky_{\mathcal{H}}$ and $h_i \notin b'$-$sky_{\mathcal{H}}$. Since $h_i \notin b'$-$sky_{\mathcal{H}}$, there must exists another object, say h_j, that b'-dominates h_i. Thus, $\forall a_r \in A : bel(h_j.a_r \geq h_i.a_r) > b'$. But, $b < b'$. Therefore, $\forall a_r \in A : bel(h_j.a_r \geq h_i.a_r) > b$. Hence, $h_j \succ_b h_i$, which leads to a contradiction as $h_i \in b$-$sky_{\mathcal{H}}$.

4.1   Experiments

We present now an extensive experimental evaluation of our approach. More specifically, we focus on two issues: (i) the size of the evidential skyline in the context of *Tripadvisor* data; and (ii) the scalability of our proposed properties for the Belief Skyline algorithm denoted by BS. We also implemented, for comparison purposes, a basic algorithm denoted by BBS (baseline belief Skyline). This later is the basic version of the BS algorithm, i.e., it does not use the properties presented in Sect. 4. The generation of the sets of evidential data is controlled by the parameters in Table 6, which lists parameters under investigation, their examined and default values. In each experimental setup, we investigate the effect of one parameter, while we set the remaining ones to their default values. The data generator and the algorithms, i.e., BS and BBS were implemented in Java, and all experiments were conducted on a 2.3 GHz Intel Core i7 processor, with 8 GB of RAM.

Table 6. Parameters and Examined Values

Parameter	Symbol	Values	Default
Number of objects	n	1 K, 2 K, 5 K, 8 K, 10 K, 50 K, 100 K, 500 K	10 K
Number of attributes	d	2, 3, 4, 5, 6	4
belief threshold	b	0.01, 0.1, 0.3, 0.5, 0.7, 0.9	0.5
Number of focal elements	f	2, 3, 5, 7, 8, 9, 10	4

Figure 3 shows the size (i.e., the number of objects returned) of the b-skyline w.r.t. n, d, b and f. Figure 3(a) shows that the size on the evidential skyline increases with higher n since when n increases more objects have chances not to be dominated. As shown in Fig. 3(b) the cardinality of the evidential skyline increases significantly with the increase of d. In fact, with the increase of d an object has better opportunity to be not dominated in all attributes. Figure 3(c) shows that the size of the evidential skyline increases with the increase of the b since the b-skyline contains the b'-dominant skyline if $b > b'$; see Property 4.

Figure 4 depicts the execution time of the implemented algorithms with regard to n, d, b and f. Overall, BS outperforms BBS. More specifically, BS is faster than BBS thanks to the properties used to improve our algorithm. As expected, Fig. 4(a) shows that the performance of the algorithms deteriorates with the increase of n. Observe that BS is one order of magnitude faster BBS since it can quickly identify if an object is dominated or not. As shown in Fig. 4(b) BBS does not scale with d. In fact, when d increases the size of the evidential skyline becomes larger. Hence, BBS performs a large number of dominance checks with a basic function. As shown in Fig. 4(c), BS is also affected by b. Figure 4(d) shows that BS is more than one order of magnitude faster than BBS.

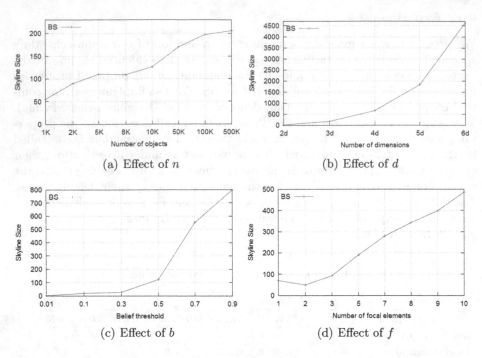

(a) Effect of n (b) Effect of d

(c) Effect of b (d) Effect of f

Fig. 3. Skyline size

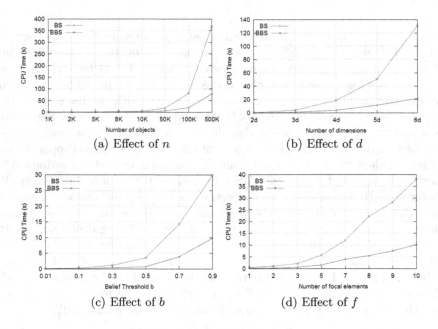

(a) Effect of n (b) Effect of d

(c) Effect of b (d) Effect of f

Fig. 4. Skyline performance

5   Conclusion and Future Works

In this paper, the *Tripadvisor* reviewers' feedbacks about hotels are treated. First, reviews are modeled as basic belief assignments. Then, the belief functions tools are used to discount and combine reviews considering the travelers' reliabilities. Since the *Tripadvisor* do not offer the multi-criteria filtering, we proposed a new Evidential Skyline operator that deals with particular type of data. The proposed Skyline operator represent an optimization of the classic skyline [7]. Finally, our skyline method is evaluated on synthetic data whose properties are similar to *Tripadvisor* ones. Experimental results are very interesting in comparison with the classic skyline method. It showed a clear optimization in terms of performance and skyline size.

Combining feedbacks when a traveler gives more than one review about a specific hotel is a promising perspective, especially that the theory of belief functions offers several combination rules for different use cases.

References

1. Bell, D.A., Guan, J.W., Lee, S.K.: Generalized union and project operations for pooling uncertain and imprecise information. Data Knowl. Eng. DKE **18**, 89–117 (1996)
2. Borzsony, S., Kossmann, D., Stocker, K.: The skyline operator. In: 17th International Conference on Data Engineering, ICDE, pp. 421–430 (2001)
3. Bousnina, F.E., Bach Tobji, M.A., Chebbah, M., Liétard, L., Ben Yaghlane, B.: A new formalism for evidential databases. In: Esposito, F., Pivert, O., Hacid, M.-S., Raś, Z.W., Ferilli, S. (eds.) ISMIS 2015. LNCS (LNAI), vol. 9384, pp. 31–40. Springer, Cham (2015). doi:10.1007/978-3-319-25252-0_4
4. Bousnina, F.E., Elmi, S., Bach Tobji, M.A., Chebbah, M., HadjAli, A., Ben Yaghlane, B.: Object-relational implementation of evidential databases. In: International Conference on Digital Economy, ICDEc, Carthage, Tunisia, 28–30 April, pp. 80–87 (2016)
5. Dempster, A.P.: Upper and lower probabilities induced by a multiple valued mapping. Ann. Math. Stat. **38**(2), 325–339 (1967)
6. Dempster, A.P.: A generalization of Bayesian inference. J. R. Stat. Soc. Ser. B **30**, 205–247 (1968)
7. Elmi, S., Benouaret, K., Hadjali, A., Bach Tobji, M.A., Ben Yaghlane, B.: Computing skyline from evidential data. In: Straccia, U., Calì, A. (eds.) SUM 2014. LNCS (LNAI), vol. 8720, pp. 148–161. Springer, Cham (2014). doi:10.1007/978-3-319-11508-5_13
8. Lee, S.K.: An extended relational database model for uncertain and imprecise information. In: 8th Conference on Very Large Data Bases, VLDB, Canada, pp. 211–220 (1992)
9. Lee, S.K.: Imprecise and uncertain information in databases: an evidential approach. In: 8th International Conference on Data Engineering, ICDE, pp. 614–621 (1992)
10. Lefevre, E., Colot, O., Vannoorenberghe, P.: Belief function combination and conflict management. Inf. Fusion **3**(2), 149–162 (2002)
11. Shafer, G.: A Mathematical Theory of Evidence. Princeton University Press, Princeton (1976)

An Adaptive Approach of Label Aggregation Using a Belief Function Framework

Lina Abassi(✉) and Imen Boukhris

LARODEC, Institut Supérieur de Gestion de Tunis,
Université de Tunis, Tunis, Tunisia
lina.abassi@gmail.com, imen.boukhris@hotmail.com

Abstract. Crowdsourcing knows a large expansion in recent years. It is widely used as a low-cost alternative to guess the true labels of training data in machine learning problems. In fact, crowdsourcing platforms such as Amazon's Mechanical Turk allow to collect from crowd workers multiple labels aggregated thereafter to infer the true label. As the workers are not always reliable, imperfect labels can occur. In this work, we propose an approach that aggregates labels using the belief function theory besides of adaptively integrating both labelers expertise and question difficulty. Experiments with real data demonstrate that our method provides better aggregation results.

Keywords: Crowdsourcing · Label aggregation · Belief function theory · Expertise · Question difficulty

1 Introduction

The phenomenon of crowdsourcing [1] has developed impressively in the last few years and represents today a very popular working technique, in particular microtasking. This latter involves human intelligence to solve simple tasks at a very cheap price. In [18], authors discovered that in some cases the workers' answers combined can be more accurate than those of a domain experts. Therefore, quality of the workers in the crowd represents the key factor of crowdsourcing. Workers can have different levels of expertise. In fact, they can be either trustworthy or bad. In the bad category, we find unskilled workers or even spammers that answer randomly in order to get more payment. Consequently, a poor quality control leads to bad results when combining answers in crowdsourcing platforms.

In this paper, we propose a new approach of label aggregation based on the belief function theory. Belief function theory offers flexibility in representing uncertainty and provides rich tools for managing different types of imperfection and combining pieces of information. Indeed, in [14], the belief function theory is applied in a crowdsourcing domain investigating an approach of experts' identification.

Our proposed approach is a solution to label aggregation problems that evaluates workers and estimates their expertise through a distance measure that

© Springer International Publishing AG 2017
R. Jallouli et al. (Eds.): ICDEc 2017, LNBIP 290, pp. 198–207, 2017.
DOI: 10.1007/978-3-319-62737-3_17

reveals the agreement level between one worker and the rest. It also suggests an estimation of question difficulty. Then, these parameters are taken into account adaptively in the aggregation process.

The remainder of the paper is structured as follows: In Sect. 2, we give an overview of the some existing works of label aggregation methods. Section 3, presents the basics of the belief function theory. Section 4 brings to light our proposed approach. Experimental evaluation is then discussed in Sects. 5 and 6 concludes our study and presents some future works.

2 Related Work

To tackle one of the most challenging problems in crowdsourcing platforms, the label aggregation [21], several methods have been proposed in literature. The easiest technique is certainly Majority Decision (MD) where the label that has the highest votes is considered as true. However considering all workers in the same level of expertise is not always optimal especially when most of them are bad. Consequently, several works focused on finding the most accurate way to aggregate labels taking into account workers reliabilities. Some of them are based on the simplest idea of testing workers with gold standard questions where true labels are available. In [11], authors propose a framework, the Expert Label Injected Crowd Estimation (ELICE), that jointly estimates worker expertise and question difficulty from workers' answers to gold standards then aggregates labels using a logistic function.

Another type of approaches propose to not rely on some gold data to estimate parameters but to do so using only the noisy labels. The Expectation Maximization (EM) proposed in [10] is an iterative method that estimates two sets of unknowns (i.e. workers expertise and true labels). Based on this latter, the Generative Model of Labels Ability and Difficulties (GLAD) [17] is a probabilistic model that also infers the question difficulty of each question. In [20], another EM-based method aggregates label and estimates worker's expertise and proposes to compute a positive expertise and a negative expertise separating positive and negative labels. Many other works [7–9,19] had also the (EM) algorithm as their basis to estimate parameters and the true labels.

In this paper, we pursue our objective to better deal with uncertainty, unlike few stated works, using the belief function theory. This latter is used in the Belief Label Aggregation (BLA) method [12] where majority decision is adopted to infer worker's accuracy but it can only be effective to a certain extent.

3 Belief Function Theory: Basic Concepts

The theory of belief function [2,3] is a theoretical framework for dealing with uncertain and partial information. This theory is a generalization of the probability and possibility theories. The Transferable Belief Model (TBM) [13] is one of several interpretations of this theory that we adopt in this work.

3.1 Representing Information

Let Ω be a finite set of elements to a given problem, called the frame of discernment. These elements are exclusive and exhaustive. All the subsets of Ω belong to its power set 2^Ω defined as follows:

$$2^\Omega = \{E : E \subseteq \Omega\}. \tag{1}$$

A basic belief assignment (*bba*) also called a mass function $m^\Omega : 2^\Omega \rightarrow [0,1]$ represents the impact of a piece of evidence on the different subsets of Ω and it is defined as follows:

$$\sum_{E \subseteq \Omega} m^\Omega(E) = 1 \tag{2}$$

Each subset E of Ω such that $m^\Omega(\mathrm{E}) > 0$ is called a focal element.

Several kinds of *bbas* are proposed expressing particular cases related to uncertainty. A *bba* is said:

- **certain**, if the only focal element E is a singleton ($m(E) = 1$).
- **vacuous**, if Ω is the only focal element ($m(\Omega) = 1$) representing the state of total ignorance.
- **categorical**, if it has a unique focal element E.
- **simple support function**, if it has at most one focal element different from Ω.

3.2 Discounting Information

Dealing with *bbas* expressed from not fully reliable sources of information requires to consider their corresponding expertise level. Therefore, discounting [2] allows to weaken *bba* by the *discount rate* $\alpha \in [0, 1]$ with $(1 - \alpha)$ is the degree of confidence of the source. The *bba* is discounted as follows:

$$\begin{cases} m^\alpha(E) = (1 - \alpha) \cdot m(E), & \forall\, E \subset \Omega, \\ m^\alpha(\Omega) = (1 - \alpha) \cdot m(\Omega) + \alpha. \end{cases} \tag{3}$$

3.3 Combining Pieces of Information

Two *bbas* m_1 and m_2 induced by distinct and reliable sources of information can be combined by several rules of combination.

Conjunctive Rule of Combination. It is introduced in [6], noted by $\bigcirc\!\!\!\cap$ and defined as follows:

$$m_1 \bigcirc\!\!\!\cap m_2(E) = \sum_{F \cap G = E} m_1(F) m_2(G) \tag{4}$$

Dempster's Rule of Combination. This rule proposed in [3] is another kind of conjunctive combination that denies the mass on the empty set. It is denoted by $\oplus$ and defined as follows:

$$\begin{cases} m_1 \oplus m_2(G) = \dfrac{m_1 \bigcirc m_2(G)}{1 - m_1 \bigcirc m_2(\varnothing)} & \text{if } E \neq \varnothing, \forall\, G \subseteq \Omega \\ 0 & \text{otherwise.} \end{cases} \tag{5}$$

The Combination with Adapted Conflict. The Combination With Adapted Confict (CWAC) rule [5] denoted by $\ominus$ is an adaptive weighting between the conjunctive and Dempster's rules acting like the first rule if *bbas* are opposite and as the second otherwise. A measure of dissimilarity allows this adaptation which is the Jousselme distance [4]. It is defined as follows:

$$d(m_1, m_2) = \sqrt{\frac{1}{2}(m_1 - m_2)^t \mathrm{D}(m_1 - m_2)}, \tag{6}$$

where D is the Jaccard index defined by:

$$\mathrm{D}(E, F) = \begin{cases} 0 & \text{if } E = F = \varnothing, \\ \dfrac{|E \cap F|}{|E \cup F|} & \forall\, E, F \in 2^{\Omega}. \end{cases} \tag{7}$$

The maximal value of all the distances can be used to find out if at least one of the sources is opposite to the others. Accordingly, the value of D can be defined as:

$$D = max[d(m_i, m_j)], \tag{8}$$

with $i \in [1, M]$ and $j \in [1, M]$ and M is the total number of mass functions. The combination rule changes to:

$$m_{\ominus}(E) = (\ominus\, m_i)(E) = Dm_{\bigcirc}(E) + (1 - D)m_{\oplus}(E) \tag{9}$$

3.4 Decision Process

In the Transferable Belief Model (TBM) [13] two levels are presented where in the first level namely the credal level, evidence is represented by mass functions and combined. As in the second level namely the pignistic level, mass functions are represented by probability functions called the pignistic probabilities denoted by $BetP$ and defined as follows:

$$BetP(\omega_i) = \sum_{E \subseteq \Omega} \frac{|E \cap \omega_i|}{|E|} \cdot \frac{m(E)}{(1 - m(\varnothing))} \quad \forall\, \omega_i \in \Omega \tag{10}$$

4   AD-BLA: Adapative Belief Label Aggregation

In this section, we detail our proposed approach namely the Adaptive Belief Label Aggregation. Our method follows two main phases. In the first, two parameters are estimated namely labeler expertise and question difficulty. In the second, these latter are integrated in the general aggregation process. We explain in more details these two big steps in what follows.

4.1   Phase 1: Parameters Estimation

Labeler Expertise Estimation. We suppose that a set of Q binary questions are given to each worker $w_j \in W$ to fulfill where the answer to each question q_i, noted a_{ij} can take as values 0, 1 or (-1) if the worker skips the question. The final aggregated label A_i exists in 0, 1 since we are dealing with binary labelling. However, we note that our method can as well manage multi-class labelling. Labeler level of expertise e_j is induced by a log distance in [0, 1] [22] that essentially computes the degree of agreement between answers of each worker and the others as follows:

$$ld(j) = -\frac{1}{n}\sum_{i=1}^{n} ln(p(q_i)) \tag{11}$$

where n is the number of questions labeled by j and $p(q_i)$ is the ratio of answers that are similar to j's. Accordingly, the worker expertise is defined as $e_j = 1 - ld_j$. Indeed, a high log distance means that the worker's answers on all the question set are far from the rest hence he has a low expertise, whereas a low log distance means that most of his answers are the same as the remaining workers hence he has a high expertise.

Question Difficulty Estimation. In this work we suppose that a worker can choose to skip a question, the question difficulty d_i is therefore estimated according to the number of labelling workers. Indeed, for each question if the number of labels with either 0 or 1 value outpasses the number of labels with (-1) value, the question is considered easy $(d_i = 0)$ as the majority chose to label it. Likewise, if the number of (-1) labels outpasses the other possibilities, the question is considered difficult $(d_i = 1)$ as the majority skipped it.

4.2   Phase 2: Aggregation Process

The first step in the aggregation process is the label modeling. Here the belief function theory is applied as it effectively represents imperfections lying within information. Accordingly, an answer is changed into a *bba* m_{ij}^{Ω} with $\Omega = \{\omega_1, \ldots, \omega_n\}$ (in our case $\Omega = \{0,1\}$).

Example 1. Let us consider the three possibilities of answers to each question q_i (0, 1 and (-1)) and transform them to *bbas*. Results are shown in Table 1.

Once the answers are modelled with *bbas*, they will be then updated using the discounting operation (Eq. 3). In fact, the discount rate of each worker (α_j) is adapted with the question difficulty regarding some conditions:

- If the question is easy and the worker has a high level of expertise (expressed by a log distance in [0, 0.2]) then his α_j is set to 0. Thus his answer is completely taken into account (the certain *bba* representing his answer stays a certain *bba*).

Table 1. Example of label modelling

Question (q_i)	$bba(i)$
0	$m_i (\{0\}) = 1$
1	$m_i (\{1\}) = 1$
(−1)	$m_i (\{\Omega\}) = 1$

– If the question is hard and the worker has a low level of expertise (ld_j in [0.8, 1]), his α_j is set to 1. This means that his answer is ignored (the certain bba becomes a vacuous bba).

In the remaining cases, the α_j takes the value of $(1 - e_j)$ referring to the log distance and $bbas$ become simple support functions after being discounted.

Example 2. We consider three workers with different discount rates and discount $bbas$ corresponding to their answers. Results are reported in Table 2.

Table 2. Example of bba discounting

Discount rate (α_j)	$bba(i,j)$	Discounted $bba(i,j)$
$\alpha_1 = 1$	$m_{i1} (\{0\}) = 1$	$m_{i1} (\{\Omega\}) = 1$ (vacuous bba)
$\alpha_2 = 0$	$m_{i2} (\{1\}) = 1$	$m_{i2} (\{1\}) = 1$ (certain bba)
$\alpha_3 = 0.3$	$m_{i3} (\{0\}) = 1$	$m_{i3} (\{0\}) = 0.7, m_{i3} (\{\Omega\}) = 0.3$ (simple support function)

Coming to the aggregation of the discounted answers of all labelers, we adopt the combination with adapted conflict (CWAC) rule (i.e. $\ominus m_i = m_{i1}^\alpha \ominus m_{i2}^\alpha \ominus ... m_{ij}^\alpha$) known as a great rule to manage conflictual information. It is even shown in [15] that this rule is more efficient than the conjunctive and Dempster's rule.

Finally, the decision process is assured by the pignistic probability (*BetP*) (Eq. 11). The final estimated label is the one having the higher pignistic probability.

Example 3. Supposing that the aggregating results for a given question q_i is the following bba:

$$m_i(\varnothing) = 0.4, m_i(\{0\}) = 0.6$$

We obtain the following pignistic probability:
BetP($\{0\}$) = 1 · (0.6/(1 − 0.4)) = 1,
BetP($\{1\}$) = 0

Accordingly, the decision is the label $A_i = 0$.

5   Experimental Evaluation

5.1   Experiment Setup

Datasets. We use the following three real-world datasets for our performance evaluation:

- The duchenne dataset [17] contains binary answers obtained through the Amazon's mechanical turk on whether images include a duchenne smile or not.
- The event temporal ordering (Temp) dataset and the recognizing textual entailment (RTE) dataset [16] are binary NPL datasets also gathered from the Amazon's mechanical turk platform.

The ground truth of all the questions in the three datasets is known in advance. A description of these datasets is given in Table 3.

Table 3. Description of datasets

Dataset	Workers (W)	Questions (Q)	Number of labels	Proportion of labels ($\neq (-1)$)
Duchenne	17	159	1221	0.45
Temp	76	462	4620	0.13
RTE	164	800	8000	0.06

5.2   Experimental Results

We compared our proposed approach (AD-BLA) to both Majority Decision (MD) and the Belief Label Aggregation (BLA) according to accuracy criteria (i.e. the proportions of correctly estimated labels) with an increasing number of labelers per question.

Figures 1, 2 and 3 plot results of our AD-BLA, MD and BLA methods as functions of the number of workers. Labelers are randomly sampled 100 times.

We observe that AD-BLA records better accuracies than the two other approaches as we increase the number of labelers. We also witness a considerable performance in the Duchenne plot particularly. It is obviously due to the impact

Table 4. Average accuracies of MD, BLA and AD-BLA for different datasets

Dataset	MD	BLA	AD-BLA
Duchenne	0.63	0.67	0.72
RTE	0.8	0.83	0.85
Temp	0.8	0.84	0.87

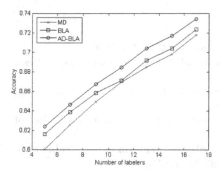

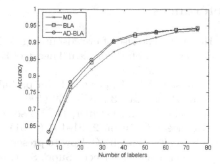

Fig. 1. Accuracies as function of workers' number for Duchenne

Fig. 2. Accuracies as function of workers' number for Temp

of the proportion of labels (different from (-1)) in our parameters estimation. The more answers are given the more accurately our approach performs but it keeps overpassing the other methods though. In Table 4 the average accuracies of these plots are presented.

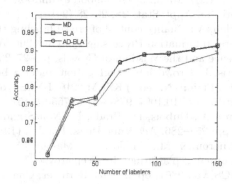

Fig. 3. Accuracies as function of workers' number for RTE

6   Conclusion and Future Works

In this work, we addressed the label aggregation problem in crowdsourcing platforms and proposed an adaptive approach of the belief label aggregation that infers the labeler expertise and question difficulty. We showed through experimentations that integrating these parameters has the ability to improve results accuracy. We manage as future works to improve even more the estimation of the two mentioned parameters and to explore other crowdsourcing scenario.

References

1. Howe, J.: The rise of crowdsourcing. Wired Magaz. **14**(6), 1–4 (2006)
2. Shafer, G.: A Mathematical Theory of Evidence, vol. 1. Princeton University Press, Princeton (1976)
3. Dempster, A.P.: Upper and lower probabilities induced by a multivalued mapping. Ann. Math. Stat. **38**, 325–339 (1967)
4. Jousselme, A.-L., Grenier, D., Bossé, É.: A new distance between two bodies of evidence. Inf. Fusion **2**, 91–101 (2001)
5. Lefèvre, E., Elouedi, Z.: How to preserve the confict as an alarm in the combination of belief functions? Decis. Supp. Syst. **56**, 326–333 (2013)
6. Smets, P.: The combination of evidence in the transferable belief model. IEEE Trans. Pattern Anal. Mach. Intell. **12**(5), 447–458 (1990)
7. Raykar, V.C., Yu, S.: Eliminating spammers and ranking annotators for crowdsourced labeling tasks. J. Mach. Learn. Res. **13**, 491–518 (2012)
8. Smyth, P., Fayyad, U., Burl, M.: Inferring ground truth from subjective labelling of venus images. In: Advances in Neural Information Processing Systems, pp. 1085–1092 (1995)
9. Yan, Y., Rosales, R., Fung, G.: Modeling annotator expertise: learning when everybody knows a bit of something. In: International Conference on Artificial Intelligence and Statistics, pp. 932–939 (2010)
10. Dawid, A.P., Skene, A.M.: Maximum likelihood estimation of observer error-rates using the EM algorithm. Appl. Stat. **28**, 20–28 (2010)
11. Khattak, F.K., Salleb, A.: Quality control of crowd labeling through expert evaluation. In: The Neural Information Processing Systems, 2nd Workshop on Computational Social Science and the Wisdom of Crowds, pp. 27–29 (2011)
12. Abassi, L., Boukhris, I.: Crowd label aggregation under a belief function framework. In: Lehner, F., Fteimi, N. (eds.) KSEM 2016. LNCS, vol. 9983, pp. 185–196. Springer, Cham (2016). doi:10.1007/978-3-319-47650-6_15
13. Smets, P., Mamdani, A., Dubois, D., Prade, H.: Non Standard Logics for Automated Reasoning, pp. 253–286. Academic Press, London (1988)
14. Ben Rjab, A., Kharoune, M., Miklos, Z., Martin, A.: Characterization of experts in crowdsourcing platforms. In: Vejnarová, J., Kratochvíl, V. (eds.) BELIEF 2016. LNCS, vol. 9861, pp. 97–104. Springer, Cham (2016). doi:10.1007/978-3-319-45559-4_10
15. Trabelsi, A., Elouedi, Z., Lefèvre, E.: Belief function combination: comparative study within the classifier fusion framework. In: Gaber, T., Hassanien, A.E., El-Bendary, N., Dey, N. (eds.) The 1st International Conference on Advanced Intelligent System and Informatics (AISI2015), November 28-30, 2015, Beni Suef, Egypt. AISC, vol. 407, pp. 425–435. Springer, Cham (2016). doi:10.1007/978-3-319-26690-9_38
16. Snow, R., et al.: Cheap and fast but is it good? Evaluation non-expert annotations for natural language tasks. In: The Conference on Empirical Methods in Natural Languages Processing, pp. 254–263 (2008)
17. Whitehill, J., Wu, T., Bergsma, J., Movellan, J.R., Ruvolo, P.L.: Whose vote should count more: optimal integration of labels from labelers of unknown expertise. In: Neural Information Processing Systems, pp. 2035–2043 (2009)
18. Alonso, O., Mizzaro, S.: Can we get rid of trec assessors? Using mechanical turk for relevance assessment. In: Proceedings of the SIGIR 2009 Workshop on the Future of IR Evaluation, vol. 15, p. 16 (2009)

19. Karger, D., Oh, S., Shah, D.: Iterative learning for reliable crowdsourcing systems. In: Neural Information Processing Systems, pp. 1953–1961 (2011)
20. Georgescu, M., Zhu, X.: Aggregation of crowdsourced labels based on worker history. In: Proceedings of the 4th International Conference on Web Intelligence, Mining and Semantics, pp. 1–11 (2014)
21. Quinn, A.J., et al.: Human computation: a survey and taxonomy of a growing field. In: Conference on Human Factors in Computing Systems, pp. 1403–1412 (2011)
22. Nicholson, B., Sheng, V.S., Zhang, J., Wang, Z., Xian, X.: Improving label accuracy by filtering low-quality workers in crowdsourcing. In: Sidorov, G., Galicia-Haro, S.N. (eds.) MICAI 2015. LNCS, vol. 9413, pp. 547–559. Springer, Cham (2015). doi:10.1007/978-3-319-27060-9_45

Assessing Items Reliability for Collaborative Filtering Within the Belief Function Framework

Raoua Abdelkhalek[(✉)], Imen Boukhris, and Zied Elouedi

LARODEC, Institut Suprieur de Gestion de Tunis,
Université de Tunis, Tunis, Tunisia
abdelkhalek_raoua@live.fr, imen.boukhris@hotmail.com
zied.elouedi@gmx.fr

Abstract. Item-based collaborative filtering is among the most widely used recommendation approaches. It consists of identifying the most similar items in order to perform recommendations accordingly. However, the reliability of the information provided by these pieces of evidence cannot be fully trusted. Hence, quantifying their reliability seems imperative to form more valuable evidence. This paper contributes to the problem of covering uncertainty in the prediction process using the belief function theory. Our approach tends to take into account the different degrees of reliability of each similar item based on the discounting factor. Then, Dempster's rule of combination is used as an aggregation operator to combine these pieces of evidence. The performance of the new evidential method is validated on a real world data set.

Keywords: Item-based collaborative filtering · Uncertain reasoning · Belief function theory · Discounting factor

1 Introduction

With the advent of Recommender Systems (RSs), an increased number of methods have been built aiming to help users finding suitable items. Among a large variety of choices, the challenge of these systems is how to generate good recommendations on which the users can rely. Actually, the collaborative filtering (CF) is the most widely implemented technique in both academia and industry [1]. Due to its great applicability, CF has been considered as a rich research area and has attracted many researchers from various fields. Owever, performances of CF are usually limited by data imperfection issues [2]. Incorporating uncertainty into the recommendation process can be argued to be an important challenge in real-world problems. Several theories have been proposed to deal with uncertainty such as belief function theory [3–5], probability theory [6] and possibility theory [7]. Among these, the belief function theory (BFT) is often presented as a generalization of the probability and the possibility theories. Due to its several advantages, the BFT has been applied in a variety of fields including RSs [8–10]. In fact, authors in [9] proposed to represent the user's preferences through

© Springer International Publishing AG 2017
R. Jallouli et al. (Eds.): ICDEc 2017, LNBIP 290, pp. 208–217, 2017.
DOI: 10.1007/978-3-319-62737-3_18

the BFT tools and integrate context information for predicting all unprovided ratings. Another method proposed in [10] relies on this theory to represent both users's preferences and community preferences extracted from social networks. An item-based collaborative filtering approach has been proposed under the BFT in [8] where the K-similar items are considered as different pieces of evidence that contribute to the final prediction. However, the reliability of information sources has not been considered. It is obvious that the most similar items should be heavily weighted in reliability measure than the other pieces of evidence. In this context, we assume in this paper that the most similar items play a more significant role than the remaining ones used in the prediction process. The main objective of this approach is to take into account both information imperfection and source reliability to form more valuable evidence and hence improving the prediction's accuracy.

The remainder of this paper is organized as follows: In Sect. 2, we present the collaborative filtering recommender. Section 3 provides a brief overview of the belief function theory. In Sect. 4, we describe our proposed recommendation approach. Then, Sect. 5 details the experimentations conducted on a real world dataset as well as a comparative evaluation. Finally, the paper is concluded and some potential future works are depicted in Sect. 6.

2 Collaborative Filtering Recommender

The collaborative filtering (CF) is the most widely used technique in the field of RSs due to its simplicity and efficiency [1]. In this section, we provide a brief overview about the CF process as well as the neighborhood-based CF.

2.1 Collaborative Filtering Process

Recommendations have always been a part of our daily life. In this context, the CF tends to provide recommendations and predict the users' interest based on historical information. It generally relies on a user-item matrix where a set of users have expressed their preferences of a set of items. Once the historical information of the users are collected, CF tries to predict their preferences and to perform recommendations accordingly. In other words, since users are not able to browse information about all available items, CF filters out irrelevant ones and suggests only items having the highest ratings.

2.2 Neighborhood-Based Collaborative Filtering

Commercial systems often rely on the so-called neighborhood-based CF methods, also referred to memory-based CF, since they are simple, easy to implement and highly effective [11]. There are two main strategies to implement a memory-based CF. The first approach is called user-based CF [12] which tends to explore what people with similar tastes have liked. The second approach is called item-based CF [13] which tends to compute similarity between items rather than users in order to recommend the potentially interesting items. Once the similarity between users or items is computed, predictions and recommendations are performed accordingly. Commonly, Pearson and Cosine correlation coefficients are the most widely used similarity measures in the neighborhood-based CF approaches [14].

3 Background on the Belief Function Theory

The belief function theory is considered as a flexible and rich framework for dealing with uncertainty [3,4]. In this section, we recall its basic concepts and operations as interpreted in the Transferable Belief Model (TBM) [5].

3.1 Frame of Discernment

In the BFT, a problem domain is represented by a finite set of elementary events called the frame of discernment, denoted by Θ, which contains hypotheses concerning the given problem [5] such that: $\Theta = \{\theta_1, \theta_2, \cdots, \theta_n\}$. In fact, all the possible values that each subset of Θ can take is called the power set of Θ and denoted by 2^Θ, where $2^\Theta = \{A : A \subseteq \Theta\}$.

3.2 Basic Belief Assignment

A basic belief assignment (bba) is an expression of the belief committed to the elements of the frame of discernment Θ [4]. It is a mapping function such that:

$$m : 2^\Theta \to [0,1] \quad and \quad \sum_{A \subseteq \Theta} m(A) = 1 \tag{1}$$

Each mass $m(A)$, called a basic belief mass (bbm), quantifies the degree of belief exactly assigned to the event A of Θ.

3.3 Dempster's Rule of Combination

Considering two bba's m_1 and m_2 induced from two reliable and independent information sources, the evidence can be combined using Dempster's rule of combination defined as:

$$(m_1 \oplus m_2)(A) = k. \sum_{B,C \subseteq \Theta : B \cap C = A} m_1(B) \cdot m_2(C) \tag{2}$$

$$where \quad (m_1 \oplus m_2)(\varnothing) = 0 \quad and \quad k^{-1} = 1 - \sum_{B,C \subseteq \Theta : B \cap C = \varnothing} m_1(B) \cdot m_2(C)$$

3.4   Discounting Operation

In most cases, the information provided by the pieces of evidence are not fully reliable. That is why, different discounting methods have been proposed in order to take into account their reliability. In this paper, we opt for that proposed by Shafer [4] where the *bba* should be altered in such a way that the obtained values are proportional to the reliability of this source as following:

$$m^\alpha(A) = (1 - \alpha) \cdot m(A), \text{ for } A \subset \Theta$$
$$m^\alpha(\Theta) = \alpha + (1 - \alpha) \cdot m(\Theta) \tag{3}$$

where $\alpha \in [0, 1]$ is a discounting factor.

4   Evidential Item-Based CF Based on Discounting Technique

Our proposed method is based on the intuition of the neighborhood-based approaches commonly used in the Recommender Systems area. As we adopt the belief function theory in order to model conveniently the uncertainty in the prediction process, we consider the K-most similar items as independent sources of evidence leading to the final prediction. The evidence collected from the selected neighbors is formalized through a basic belief assignment (*bba*). Then, a reliability measure is incorporated into the mass function using a discounting technique. Finally, the discounted *bba*'s are combined using Dempster's rule of combination and the final *bba* corresponding to the target item is provided to the active user. The whole process of our new recommendation approach is illustrated in Fig. 1.

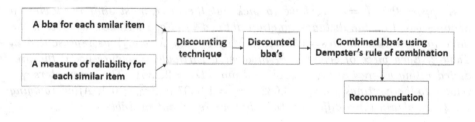

Fig. 1. A new prediction process using the discounting technique

4.1   Evidence Representation

In our approach, the rating provided by each similar item will be transformed into a mass function $m(\{\omega_i\})$ based on the formalism proposed in [15]. Note that $\Theta = \{\omega_1, \omega_2, \cdots, \omega_n\}$ where n is the number of the possible ratings ω that can be provided by the users. We can represent this *bba* as following:

$$m_{x,y}(\{\omega_i\}) = \alpha_0 \exp^{-(\gamma_{\omega_i}^2 \times (D(x,y)/max(D))^2)}$$
$$m_{x,y}(\Theta) = 1 - \alpha_0 \exp^{-(\gamma_{\omega_i}^2 \times (D(x,y)/max(D))^2)} \tag{4}$$

where α_0 is fixed to the value 0.95 as invoked in [15], γ_{ω_i} is the inverse of the mean distance between each pair of items sharing the same rating value ω_i and $D(x, y)$ is the distance between the item x and the item y defined as:

$$D(x, y) = \frac{\sqrt{\sum_{u \in u_x \cap u_y} (r_{u,x} - r_{u,y})^2}}{|u_x \cap u_y|} \tag{5}$$

Where $r_{u,x}$ and $r_{u,y}$ are the ratings given by user u for the items x and y, u_x and u_y are the users who have rated respectively the items x and y and $max(D)$ is the maximum value over all the computed distances.

Example 1. *Based on the user-item matrix illustrated in Table 1, let us consider $User_5$ as the active user and $Movie_7$ as the target movie.*

Table 1. User-item matrix

	$Movie_1$	$Movie_2$	$Movie_3$	$Movie_4$	$Movie_5$	$Movie_6$	**Movie7**
$User_1$	2	4	?	3	3	5	3
$User_2$	5	5	2	4	2	4	?
$User_3$	?	?	4	2	4	1	5
$User_4$	3	1	5	1	?	?	2
User5	1	5	5	3	3	1	?

In this example, the scale rating ranges from to 1 to 5. The frame of discernment corresponding to this situation is then: $\Theta = \{1, 2, \cdots, 5\}$.

Suppose that $K = 3$, we have to pick out the three most similar movies to $Movie_7$ based on the distance function defined in (Eq. 5). Hence, we obtain the three most similar movies $= \{Movie_5, Movie_1 \text{ and } Movie_2\}$ as represented in Table 2. Each piece of evidence P_i involves a particular evidence about the predicted rating where $i = \{1, \cdots, K\}$ and $max(D) = 2.2361$. The parameters γ_{ω_i} related to their ratings are $\gamma_1 = 0.6325$, $\gamma_2 = 1.0607$ and $\gamma_5 = 0.4$. After applying Eq. 4, we obtain three different bba's that we represent in Table 3.

Table 2. Distances between the target movie and the other movies

	$Movie_1$	$Movie_2$	$Movie_3$	$Movie_4$	$Movie_5$	$Movie_6$
$Movie_7$	**0.7071**	**0.7071**	1.5811	1.0541	**0.5**	2.2361

4.2   Reliability Evaluation

Although the *bba's* are generated based on the computed distances, the two parameters γ and α affect the prediction results. Since the similar items' ratings

Table 3. The *bba*'s related to the pieces of evidence

Pieces of evidence	Corresponding *bba*'s
P_1 : $Movie_5$	$m_{Movie_7,Movie_5}(\{3\}) = 0.7387$ *and* $m_{Movie_7,Movie_5}(\Theta) = 0.2613$
P_2 : $Movie_1$	$m_{Movie_7,Movie_1}(\{1\}) = 0.8371$ *and* $m_{Movie_7,Movie_1}(\Theta) = 0.1629$
P_3 : $Movie_2$	$m_{Movie_7,Movie_2}(\{5\}) = 0.9031$ *and* $m_{Movie_7,Movie_2}(\Theta) = 0.0969$

are not fully reliable, we propose in this phase to apply the discounting operation [4] in order to quantify their degree of reliability. We divide the obtained distance for each similar item by the maximum value in order to get normalized values in [0,1] to be used in the reliability computation. Hence, we obtain the discounting factor that we denote by β, where $\beta = \frac{D(x,y)}{max(D)}$. This phase consists of considering the reliability of each information source. In our case, we assume that the more similar the item is, the more reliable its evidence is. Therefore, the idea is to weight most heavily the evidence of the items having the lowest distances and conversely for the less reliable ones. Note that the smaller the discounting, the higher the reliability. Thus, $m_{x,y}$ is discounted as follows:

$$m_{x,y}^{\beta}(\{\omega_i\}) = (1 - \beta) \cdot m_{x,y}(\{\omega_i\})$$
$$m_{x,y}^{\beta}(\Theta) = \beta + (1 - \beta) \cdot m_{x,y}(\Theta)$$

(6)

Example 2. *Based on the computed distances (Table 2), we aim to discount the bba of each similar movie in order to quantify its degree of reliability. As pointed out previously, we denote $\beta = \frac{D(x,y)}{max(D)}$ the discount rate. This strategy consists in discounting the three pieces of evidence P_1, P_2 and P_3: $\beta_1 = \frac{0.5}{2.2361} = 0.2236, \beta_2 = \frac{0.7071}{2.2361} = 0.3162$ and $\beta_3 = \frac{0.7071}{2.2361} = 0.3162$. P_1 is supposed to be the most reliable one since $\beta_1 < \beta_2$ and $\beta_1 < \beta_3$. The discounted bba's are depicted in Table 4.*

Table 4. The discounted *bba*'s related to the pieces of evidence

Pieces of evidence	Discounted *bba*'s
P_1 : $Movie_5$	$m_{Movie_7,Movie_5}^{\beta}(\{3\}) = (1\text{-}0.2236).0.7387 = 0.5735$
	$m_{Movie_7,Movie_5}^{\beta}(\Theta) = 0.2236 + (1\text{-}0.2236).0.2613 = 0.4265$
P_2 : $Movie_1$	$m_{Movie_7,Movie_1}^{\beta}(\{1\}) = (1\text{-}0.3162).0.8371 = 0.5724$
	$m_{Movie_7,Movie_1}^{\beta}(\Theta) = 0.3162 + (1\text{-}0.3162).0.1629 = 0.4276$
P_3 : $Movie_2$	$m_{Movie_7,Movie_2}^{\beta}(\{5\}) = (1\text{-}0.3162).0.9031 = 0.6175$
	$m_{Movie_7,Movie_2}^{\beta}(\Theta) = 0.3162 + (1\text{-}0.3162).0.0969 = 0.3825$

4.3   Fusion of Pieces of evidence

Once the *bba*'s are discounted, Dempster's rule of combination is used in the aggregation process (Eq. 2) and the target item is recommended to the user.

Example 3. *The rating provided by $User_5$ to $Movie_7$ is obtained by the aggregation of the three bba's already discounted in Example 2. Hence, the recommended movie has the following bba's: $m(\{1\}) = 0.1539$, $m(\{3\}) = 0.0938$, $m(\{5\}) = 0.2872$ and $m(\Theta) = 0.4651$. In other words, the prediction given to $User_5$ indicates for him that a bbm of 0.2872 supports that he will extremely like $Movie_7$. The evidence that he will never be interested in such movie has a value of 0.1539. A moderate satisfaction towards the suggested movie is around 0.0938 while the rest of the committed belief is allocated to the frame of discernment Θ.*

5   Experiments and Discussions

To evaluate our approach, we resort to one of the widely used real world data set in CF which is publicly available on the MovieLens[1] website. In the MovieLens data set, 943 users have rated 1682 movies leading to 100.000 ratings. We followed the method suggested in [16] for conducting our experiments. The movies rated by the 943 users are ranked according to the number of the ratings given by the users. The experimentation protocol consists on increasing progressively the number of the missing rates. Hence, we obtain 10 different subsets containing a specific number of ratings provided by the 943 users for 20 different movies in the data set. For each subset, we randomly extract 20% of the available ratings as a testing data and the remaining 80% were considered as a training data.

5.1   Evaluation Metrics

During our experiments, we opt to two evaluation metrics commonly used in CF: the *Mean Absolute Error* (MAE), $\in [0, 4]$ in this case, in order to assess the prediction's accuracy as well as the precision measure, $\in [0, 1]$, in order to evaluate the quality of recommendations provided to the active user such that:

$$MAE = \frac{\sum_{u,i} |\widehat{R}_{u,i} - R_{u,i}|}{\|\widehat{R}_{u,i}\|} \tag{7}$$

$$Precision = \frac{IR}{IR + UR} \tag{8}$$

where $R_{u,i}$ is the real rating for the user u on the item i and $\widehat{R}_{u,i}$ is the predicted value. $\|\widehat{R}_{u,i}\|$ is the total number of the predicted ratings over all the users. IR indicates that an interesting item has been correctly recommended to the active user while UR indicates that an uninteresting item has been incorrectly recommended to the active user. Note that the lower values of MAE mean a better prediction's accuracy while the higher the precision measure is the more effective the recommendation approach is.

---

[1]   http://movielens.org.

5.2 Results

We carry on several experiments over the 10 subsets by varying each time the neighborhood size K from 1 to 10. For each subset, we compute the MAE and the precision measure for each value of K and we note the overall results. Unlike the evidential item-based CF (EV-IBCF) [8], the proposed approach considers the reliability of the information sources and aggregates the discounted *bba*'s leading to more valuable evidence. Hence, we compare it to our approach which is based on a discounting technique (DE-IBCF). Besides, we implement both Pearson (P-IBCF) and Cosine-based approaches (C-IBCF) [14] and we compare them with our method. Table 5 recapitulates results considering different sparsity degrees.

Table 5. Overall MAE and precision

Evaluation metrics	Subsets	Sparsity degrees	EV-IBCF	C-IBCF	P-IBCF	DE-IBCF
MAE	$Subset_1$	53%	0.751	0.824	0.839	0.711
Precision			0,79	0,778	0,774	0,774
MAE	$Subset_2$	56.83%	0.84	0.87	0.9361	0.802
Precision			0,76	0,739	0,737	0,748
MAE	$Subset_3$	59.8%	0.761	0.825	0.863	0,836
Precision			0,77	0,749	0,752	0.711
MAE	$Subset_4$	62.7%	0.763	0.876	0.905	0.743
Precision			0,763	0,745	0,746	0,775
MAE	$Subset_5$	68.72%	0.831	1	0.990	0.802
Precision			0,741	0.69	0,707	0,787
MAE	$Subset_6$	72.5%	0.851	0.917	0.976	0.843
Precision			0,735	0,733	0,732	0,74
MAE	$Subset_7$	75%	0.744	0.877	0.943	0.736
Precision			0,78	0,745	0,752	0,783
MAE	$Subset_8$	80.8%	0.718	0.848	0.927	0.723
Precision			0,778	0.718	0,729	0,821
MAE	$Subset_9$	87.4%	0.840	0.978	0.958	0.839
Precision			0,707	0,6539	0,665	0,74
MAE	$Subset_{10}$	95.9%	0.991	1.13	0.913	0.978
Precision			0,513	0,509	0,463	0,431
Overall MAE			0.809	0.914	0.925	**0.789**
Overall Precision			0,733	0,706	0,706	**0.743**

In terms of MAE, DE-IBCF achieves better results corresponding to a value of 0.789 compared to EV-IBCF, C-IBCF and P-IBCF having respectively 0.809, 0.914 and 0.925. Besides, the results related to the precision measure indicates that the two correlation-based CF have almost the same behavior (0.706) while the DE-IBCF performs significantly better than the three other approaches (0.743 compared to 0.706 and 0.733). Thereupon, the empirical results show that using the discounting method allows an improvement over the standard item-based CF approaches by acquiring the lowest error rates as well as the highest overall precision. Aiming to better emphasize the performance of our approach, we resort also to the paired t-test [17] which reports whether the mean of the differences over each pair of methods is statistically significant or not for both MAE and precision. For instance, using paired t-tests for the precision criterion, the DE-IBCF achieves better results than the P-IBCF and C-IBCF with the same p-value = 0.0008. Compared to EV-IBCF, the DE-IBCF has consistently better precision with p-value = 0.027. Consequently, these results indicate that the improvement of DE-IBCF is statistically significant and hence more efficient.

The performance of the three recommendation approach is also investigated considering different values of K as illustrated in Figs. 2 and 3.

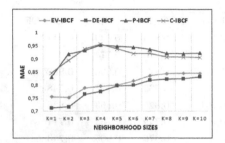

Fig. 2. Overall MAE versus K   **Fig. 3.** Overall precision versus K

We observe that the size of neighborhood affects the performance of the different approaches. In fact, we note that the error values increase as we increase the neighborhood size from 1 to 7. After this value of K, the values of MAE corresponding to all the methods are almost constant. However, DE-IBCF achieves the lowest error rates over the different values of the parameter K. For the precision measure, we have almost the opposite behaviors. When varying the value of K, the curve of DE-IBCF remains overhead with an optimum value corresponding to 0.793 when K = 1.

6   Conclusion

In this paper, we have proposed an item-based collaborative filtering approach that uses the belief function theory for representing, discounting and combining evidence provided by the K-similar items. Since we cannot admit that these pieces of evidence are fully trusted, we have adopted a discounting technique to quantify the reliability given to each similar item. Integrating a reliability measure into the recommendation process improves the prediction's accuracy which is experimentally proved. A future direction is to take the ratings' distribution into account. Thus, users' ratings should be weighted differently in distance measure. We intend also to use an other combination rule in the aggregation step.

References

1. Park, Y., Park, S., Jung, W., Lee, S.G.: Reversed CF: a fast collaborative filtering algorithm using a k-nearest neighbor graph. Expert Syst. Appl. **42**(8), 4022–4028 (2015)
2. Nguyen, V.-D., Huynh, V.-N.: A community-based collaborative filtering system dealing with sparsity problem and data imperfections. In: Pham, D.-N., Park, S.-B. (eds.) PRICAI 2014. LNCS, vol. 8862, pp. 884–890. Springer, Cham (2014). doi:10. 1007/978-3-319-13560-1_74
3. Dempster, A.P.: A generalization of bayesian inference. J. Roy. Stat. Soc. Ser. B (Methodol.) **30**, 205–247 (1968)
4. Shafer, G.: A Mathematical Theory of Evidence, vol. 1. Princeton University Press, Princeton (1976)
5. Smets, P.: The transferable belief model for quantified belief representation. In: Smets, P. (ed.) Quantified Representation of Uncertainty and Imprecision, pp. 267–301. Springer, Dordrecht (1998)
6. Chow, Y.S., Teicher, H.: Probability Theory: Independence, Interchangeability, Martingales. Springer Science and Business Media, New York (2012)
7. Dubois, D., Prade, H.: Possibility Theory and Its Applications: Where Do We Stand?. Springer Handbook of Computational Intelligence. Springer, Berlin Heidelberg (2015)
8. Abdelkhalek, R., Boukhris, I., Elouedi, Z.: Evidential item-based collaborative filtering. In: Lehner, F., Fteimi, N. (eds.) KSEM 2016. LNCS, vol. 9983, pp. 628–639. Springer, Cham (2016). doi:10.1007/978-3-319-47650-6_49
9. Nguyen, V.-D., Huynh, V.-N.: A reliably weighted collaborative filtering system. In: Destercke, S., Denoeux, T. (eds.) ECSQARU 2015. LNCS, vol. 9161, pp. 429–439. Springer, Cham (2015). doi:10.1007/978-3-319-20807-7_39
10. Nguyen, V.-D., Huynh, V.-N.: Integrating with social network to enhance recommender system based-on dempster-shafer theory. In: Nguyen, H.T.T., Snasel, V. (eds.) CSoNet 2016. LNCS, vol. 9795, pp. 170–181. Springer, Cham (2016). doi:10. 1007/978-3-319-42345-6_15
11. Su, X., Khoshgoftaar, T.M.: A survey of collaborative filtering techniques. Adv. Artif. Intell. **4**, 1–19 (2009)
12. Zhao, Z.D., Shang, M.S.: User-based collaborative-filtering recommendation algorithms on hadoop. In: International Conference on Knowledge Discovery and Data Mining, pp. 478–481 (2010)
13. Kim, H.-N., Ji, A.-T., Jo, G.-S.: Enhanced prediction algorithm for item-based collaborative filtering recommendation. In: Bauknecht, K., Pröll, B., Werthner, H. (eds.) EC-Web 2006. LNCS, vol. 4082, pp. 41–50. Springer, Heidelberg (2006). doi:10.1007/11823865_5
14. Aggarwal, C.C.: Neighborhood-Based Collaborative Filtering. In: Recommender Systems, pp. 29–70. Springer International Publishing (2016)
15. Denoeux, T.: A K-nearest neighbor classification rule based on Dempster-Shafer theory. IEEE Trans. Syst. Man Cybernet. **25**(5), 804–813 (1995)
16. Su, X., Khoshgoftaar, T.M.: Collaborative filtering for multi-class data using bayesian networks. Int. J. Artif. Intell. Tools **17**(01), 71–85 (2008)
17. Pandis, N.: Comparison of 2 means for matched observations (paired t test) and t test assumptions. Am. J. Orthod. Dentofac. Orthop. **148**(3), 515–516 (2015)

Author Index

Printed in the United States
By Bookmasters